AF180373

BIOGEOGRAPHY AND BIODIVERSITY

Judith Rosales

SOCIETY Publishing

www.societypublishing.com

Biogeography and Biodiversity

Judith Rosales

Society Publishing
2010 Winston Park Drive,
2nd Floor
Oakville, ON L6H 5R7
Canada
www.societypublishing.com
Tel: 001-289-291-7705
001-905-616-2116
Fax: 001-289-291-7601
Email: orders@arclereducation.com

© 2019 Society Publishing

ISBN: 978-1-77361-206-5 (Hardcover)

This book contains information obtained from highly regarded resources. Reprinted material sources are indicated and copyright remains with the original owners. Copyright for images and other graphics remains with the original owners as indicated. A Wide variety of references are listed. Reasonable efforts have been made to publish reliable data. Authors or Editors or Publishers are not responsible for the accuracy of the information in the published chapters or consequences of their use. The publisher assumes no responsibility for any damage or grievance to the persons or property arising out of the use of any materials, instructions, methods or thoughts in the book. The authors or editors and the publisher have attempted to trace the copyright holders of all material reproduced in this publication and apologize to copyright holders if permission has not been obtained. If any copyright holder has not been acknowledged, please write to us so we may rectify.

Notice: Registered trademark of products or corporate names are used only for explanation and identification without intent of infringement.

Society Publishing publishes wide variety of books and eBooks. For more information about Society Publishing and its products, visit our website at www.societypublishing.com.

ABOUT THE AUTHOR

Prof. Judith is an Ecologist and Environmental Specialist with more than 30 years of academic and consulting experience. She holds a PhD in geography from the University of Birmingham UK and wrote the PhD and Master Academic Programs of Environmental Sciences for the Universidad de Guayana, Venezuela.

TABLE OF CONTENTS

LIST OF FIGURES

LIST OF TABLES

PREFACE

A part of the universe that was formed 4.6 billion years ago is still a very interesting topic to explore despite its age. Yes, it's the earth that we are talking about. Scientific literature has never failed anybody's curiosity, and it never will for the years to come. This book intended for scholars at the undergraduate level focuses on a unique approach to the study of all the organisms living in different habitats spread across the earth's land and the waters. The geography of the earth is considered as a set of bioregions that abound in particular taxa, landscape features and climatic factors.

The diversity in every region has interesting insights on the evolution and the history of earth. The vastness of this topic is a measure that is unfathomable when considered from the research point of view. There is no limit to the studies that can be conducted and are dedicated to an elaborated study of the existence of ecosystems and their respective interactions. Various mathematical tools were also used to arrive at concrete conclusion of several aspects when studying a particular range of species.

This book is an attempt to identify general patterns of biogeography and biodiversity on the basis of well-developed natural environments and their transformation over time due to the climatic and topographical changes. This is also signified by a subject already published in an earlier book of this environmental sciences series as it is the movement of tectonic plates over the years ultimately leading to the creation of more accommodating niches suiting to the flourishment of biodiversity. This book also dives into the process of the encroachment into nature by humans and its subsequent concerns with reference to the effects on the existence of the species and their conservation.

With the objective of keeping a sequential approach to this study that is easy to follow, the subject matter of this book begins with the biogeography that focuses on the earth's major regions. The next section will seek biodiversity at different scales and levels as to how the species abundance of certain areas made provisions for the productivity of that ecosystem. The interdependence of the resources from an ecosystem with the economy is taken next and explored with reference to the current context of nature degradation beyond admissible levels and its effects on climate change, which is also a subject of another earlier book in this environmental science series.

The depreciation of biodiversity and its necessary restoration is a hot topic that has taken multiple levels in terms of research and the development of processes contributing to the revamping of the affected habitats. The fate of biodiversity is already showing clear demonstrations of how it has been affected with the development of the countries over time. A bird's eye view of the flora and fauna of these regions, existing theories explaining species diversity, are all covered next followed by a thorough detail of the ongoing conservation practices. Up next are the recent developments in the research of biogeography with the assistance of improved methodologies and modern technologies. The last section has been specially designed to provide leads for further research has found relevant to the topics in consideration.

1
CHAPTER

INTRODUCTION TO BIOGEOGRAPHY

"Of these three essential factors, SPACE might be said to be one with which biogeography is primarily concerned. However, space necessarily interplays with TIME and FORM; therefore, the three factors are as one of biogeographic concern."

—Leon Croizat

CONTENTS

ABSTRACT

"Biogeography is an integrative field of inquiry that unites concepts and information from ecology, evolutionary biology, geology, and physical geography." Biogeography refers to the circulation of species and ecosystems in geographic space and through geological time. In this chapter, we will talk about the basic concepts of biogeography and how it changes with the changing world. In the second half of the chapter, the book refers biogeography in 18th, 19th, and 20th centuries, and different thoughts and proposals of various scientists that were important for that time are discussed. At the end of the chapter, the rise of the thought of vicarianism and the beginning of a new era in biogeography are presented.

1.1 INTRODUCTION

Biogeography is an interdisciplinary science that studies the distribution of living beings in space and through the time. The aim is to study the distribution of life based on its dynamics on the spatial and time scales on Planet Earth. Specific goals of biogeography are the understanding of:

- how organisms are distributed on the planet, and
- where the organisms are located and why the organisms are in a specific place?

It means to study distribution patterns of species.

In biology, it enters within the field of comparative biology that grew up through the accumulation of data on the diversity of the attributes of organisms. Because of these studies, they were described and named to more than one and a half million species. However, the mere accumulation of data on the world we live in, no matter how well ordered they are, is not enough to understand that world. The theory of evolution, formulated by Charles Darwin and Alfred Russel Wallace in the mid-19th century, allows us to interpret the data accumulated by the comparative biology.

The study of the diversity of life, now looked through the lens of the theory of evolution, is equivalent to the study of the history of life. Therefore, the analysis of the form (similarities and differences in the attributes of organisms) can no longer be separated from the consideration of time and space. Darwin first referred to the continental drift theory in which all the continents were grouped together into the middle of the Pacific Ocean. Later, in his book called "On the Origin of Species," he changed his mind and thought about continents and oceans being together. However, he is the first to nominate the science of biogeography founded in some geological theories.

Biogeography can be studied with a focus on the historical factors that have shaped the current distributions, or with a focus on ecological factors that shape the current distribution of organisms. Zoogeography and phytogeography are the two subdisciplines of biogeography, which respectively involve the distribution of plants and animals. The taxonomic groups in the ecosystem vary geographically not only based on major environmental factors of the habitats they occupy, like latitude and elevation but also based on the degree of isolation that normally relates to genetic specialization.

The three main areas of biogeography are: historical biogeography, ecological biogeography, and conservation biogeography. In all the three areas, the species distribution is addressed from a very different perspective as discussed below:

- **Ecological biogeography:** It studies how the ecological processes that occur in the short run act on the distribution of organisms and analyzes the distribution of living beings according to their adaptations to the current conditions of the medium. In ecological biogeography, the global distribution of plant and animal species is involved. The most common contributing factors or the ecological factors that are studied commonly under ecological biogeography are habitat, climate, and primary productivity (the rate at which plants in a particular ecosystem produce the net chemical energy).
- **Historical biogeography:** It studies how ecological processes that occur in the long run act on the pattern of distribution of organisms and explains the distribution of living beings in function of historical factors. Therefore, historical biogeography is seen from the evolutionary perspective over time involving phylogenetic distributions.

There are a number of factors that have influenced and influence how organisms are distributed on the planet: plate tectonics, uprights, narrowing/widening of the bed of a river, glaciations, vegetation physiognomies, climate, humidity, marine currents, and so forth. Therefore, it is extremely difficult to establish a dividing line between what would be an "ecological" or "historical" factor. That is why some authors refute the existence of a historical biogeography. As "ecology" and "history" have played roles side by side since always, it is obvious that they are inextricably tied to one another.

Through the comparison of similar species with minor differences that originated due to adaptations to their respective environments, biogeography provides evidence of evolution. The creation of novel climates and habitats are the result of Earth's continental separation, drifted apart, and colliding. As species modifies themselves in response to the present conditions, members of the same species that had been separated geographically diverge, resulting in the eventual formation of distinct species. It is very important to have such knowledge, as it helps in understanding how adaptations occurred in response to changing environments in the past, and how this knowledge can be used for future conservation purposes. Present environmental conditions are related to the distribution of species in large biomes and ecoregions along the world.

For the reader having an interest in the ecological knowledge different scales, it is advisable to read the general ecology book of Eugene Odum and Barrett Gary, where there is a thoroughly description of seven ecological levels—organism, population, community, ecosystem, landscape, biome, and ecosphere—in terms of seven transcending functions—energetics, evolution, development, regulation, behavior, diversity, and integration (see Bibliography).

1.2 ECOLOGICAL BIOGEOGRAPHY

Ecological biogeography studies the factors that define the present spatial distribution of organisms. In ecology, the spatial distribution of organisms relates to environmental factors that influencc the species niche or competitive fitness of organisms in populations or species in ecosystems. The theoretical framework includes ecological theories, such as the islands theory, that were the basis for the theory of evolution. The islands theory today is analyzed not only with real islands in the sea but also within local geographical spaces with functional islands; the theory of niche as a super volume of environmental variables within a species finds its ecological optimums, the meta-population concept that includes how a population can be found along a regional space without becoming extinguished, the meta-communities concept including regional spaces for location of communities and finally the neutral theory of biogeography and biodiversity that states that location of individuals is driven by random events.

The basic ecological science has the environmental knowledge about what determines the distribution of living beings at a regional scale: being climatic variables the first of them.

The main factor to differentiate the distribution of global biomes is its terrestrial or aquatic environmental type and from the last marine or freshwater characteristics of the water body as salinity is a determinant in the distribution of all aquatic organisms.

Water as all inside and outside cells of living organisms: Mammals, for instance, have more than 70% of water by weight, 46% inside cells, and about 23% outside cells in blood plasma and other body fluids. Virtually all chemical reactions in life processes take place in solution in water. Some organisms can live in a dormant and desiccated state for long durations but require water to become active.

Temperature. The metabolism of living beings requires temperatures within very narrow margins precise. Excessive cold precludes exchanges between soil and plants, stops chlorophyll assimilation, considerably slows down respiration, and deteriorates organs or tissues that are more exposed. On the other hand, temperatures that are too high denaturalize enzymes on which the metabolism depends and damages or destroys molecular structures and cell membranes. Because of that, life is very difficult below −10°C and or above 50°C.

Energy. Virtually all living beings depend directly or indirectly on light, and that this is what makes the chlorophyll function possible through which vegetables synthesize organic matter from mineral substances. The only exceptions to this rule are found among some very specialized or extremophiles like the bacteria living underground or in the extraordinary ecosystems of caves or oceanic pits that take advantage of the heat terrestrial and sulfurous emissions.

Also, wind is one of the variables that present regional variation and help in the distribution of organisms and presence of different ecosystems and biomes in present.

Those great climatic factors described so far have an influence at scale regional for the distribution of species and biomes. They are responsible for the latitudinal variations along the earth. The climates artic, temperate, tropical, xeric are the factors behind distribution of species and ecosystem.

There are also other factors and elements capable of altering the ecological distribution at local scales, and they are more related to the distribution of ecosystems and communities:

- Geomorphology includes altitude, exposure, and slope (features that determine the topoclimates, runoff, soil formation, or other essential factors for living beings). The link between geomorphological factors and ecological factors is essential to explain local variations end environmental gradients as geomorphological processes depend on geology and climate; they all together forms the basis for the spatial variation of vegetation communities. Geomorphology also influences and is influenced by hydrological factors, which, in turn, determines gradients in aquatic marine or freshwater alluvial ecosystems.
- Edaphic (dependent on the climate and the geologic substrate) of which it is going to depend on the existence of more or less productive soils, acidic, or alkaline or with the presence of some or other substances capable of influencing the vegetation.
- Interactions with other species such as herbivory, pollination, and dispersion. Biological interactions within each ecosystem, organisms and populations maintain a complex set of relationships with other individuals of their own species or with those of the rest. These interactions, which depending on the cases may be favorable, unfavorable, or indifferent to the survival of the affected species, can determine their distribution area while playing an essential role in their evolution (e.g., dams "learn" to hide or defend themselves from their predators that, in turn, will develop more effective hunting techniques and so on, successively). Intraspecific relationships (among individuals of the same species) condition the behavior and the territory occupied by a population but do not threaten the continuity of the species. In the case of animals, they determine the solitary character or gregariousness of individuals and are often related to reproduction, feeding or protection of the offspring.
- Human use: Also depending on human use, species can be moved from its origin.

This is not a treaty of ecology; however, biogeography, as a combination of biology and geography at large scales in geographical ranges, is based on basic ecological principles of the ecological theory. Therefore, we need to know at least what are terrestrial and aquatic global biomes.

1.2.1 Terrestrial Biomes

These are the major biomes found in continental lands.

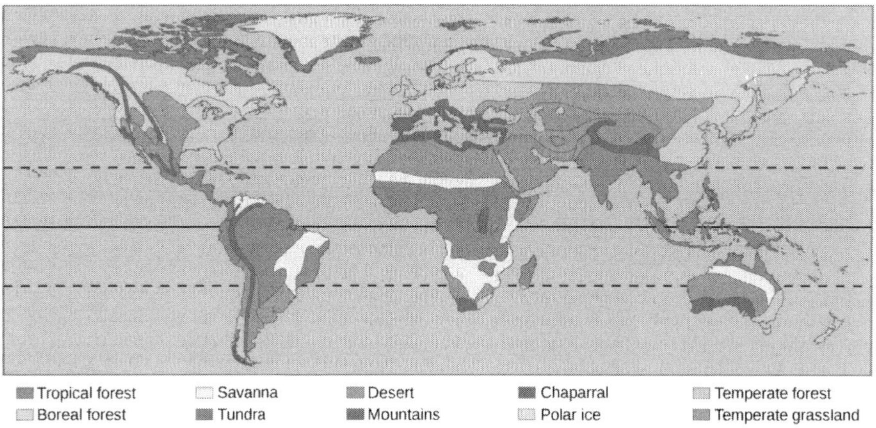

Figure 1.1: Biomes on earth. Source: https://openoregon.pressbooks.pub/envirobiology/

General description of terrestrial biomes:

1.2.1.1 Tropical Forests

Tropical forest is the world's richest forest in terms of species, having up to four layers in their vertical structure and can reach a maximum altitude of 50 m. They areclassified into tropical rain {>40 m (very tall) and 25–40 m (tall)}, tropical deciduous {15–25 m (medium)}, and tropical dry {10–12 m (short)} forests. They have a complex vertical structure with layers of understory, lower strata, middle strata, upper strata, and emergent strata representing the layering of the forest with many distinct habitats for animals within each layer.

Figure 1.2: Mora forests and a view of the rain forests in Iwokrama, Guyana.

Photographs from Dr. Raquel Thomas

1.2.1.2 Boreal Forests or Taiga

Boreal forest is the biome characterized by coniferous forests consisting mostly of pines, spruces, and larches, and is localized in boreal environment. They are mainly located between latitudes 50° and 60° N In that region there are short, cool summers and long, cold winters. Low annual temperatures characterize the environment of the boreal forest/or taiga (i.e., the swampy coniferous forest of high northern latitudes, especially that between the tundra and steppes of Siberia.. Precipitation ranges from 40 to 100 cm per year.

Figure 1.3: Boreal Forest in Finland (Photograph from Antti Mikkola).

1.2.1.3 Savannas

Savannas are areas of open grassland with or without being spreadto dense trees cover. They are present in tropical regions where climates are characterized by a seasonal rainfall with a dry period followed by a humid period. Fire is one of the factors that maintain the savannas' biome. Palinnologists have demonstrated that savanna floristic composition can be dated in each change from glacial to interglacial humid periods of the quaternary. Dominant species in the herbaceous component of savannas are species from the Poaceae family.

Figure 1.4: The savannas from North America in Gran Sabana, Venezuela

(Photograph from Evelyn Guzman)

1.2.1.4 Tundras

Tundra is the coldest of all biomes and means a plain without trees. Here, the weather is very cold with very short summers. Temperatures can reach −45.5°C. Vegetation includes mosses, lichens, and small shrubs.

Figure 1.5: Tundra biome. Photograph from https://www.thinglink.com/scene/495423976990834689

1.2.1.5 Deserts

Deserts are the biomes characterized by minimum rainfall of <500 mm. Temperatures are extreme here: 49°C in summers and 2–4°C in winters. However, they also show their extremes, as it gets extremely cold at night and hot during day time.

Figure 1.6: Desert biome (Photograph from https://www.quotemaster.org/Desert).

1.2.1.6 Chaparral

This biome is characterized by semi-arid, shrub-dominated, hard-leaved, woody plants within Mediterranean type climates (wet winters, and hot and dry summers).

Figure 1.7: The Santa Monica chaparral.

https://www.nps.gov/fire/wildland-fire/resources/images/santa-monica-chapparal-view.jpg

1.2.1.7 Polar ice

The polar ice biome is the coldest place on the earth; temperatures are usually below 0°C. In summer, the sun could shine continuously for 24 hours.

Figure 1.8: Polar ice.

Blue Planet http://www.telegraph.co.uk/content/dam/Travel/2017/October/blue-planet-walrus-small.jpg)

1.2.1.8 Temperate forest

These are forests that occur between 30 and 50° of North Latitude. They have a continuous cover and can reach up to 40 m tall with three strata.

Figure 1.9: Temperate forest in Columbia County (http://hvfarmscape.org/).

1.2.1.9 Temperate grasslands

Temperate natural grasslands are ecosystems with up to 50 plant species dominated by grasses and sedges developed in regions characterized by an annual rainfall between 250 and 750 mm (10 and 30 in), a high rate of evaporation, and seasonal and annual droughts.

Figure 1.10: Beautiful grasslands in a British landscape. Source from tourists:

(https://i.pinimg.com/736x/fc/ec/df/fcecdf2936a566cf7924682952c00c9c--england-uk-travel-england.jpg).

1.2.2 Aquatic Biomes

Aquatic biomes are those where aquatic habitats house all processes of living beings. They can also be differentiated by marine or freshwater and the average temperature that follow the latitudinal zones.

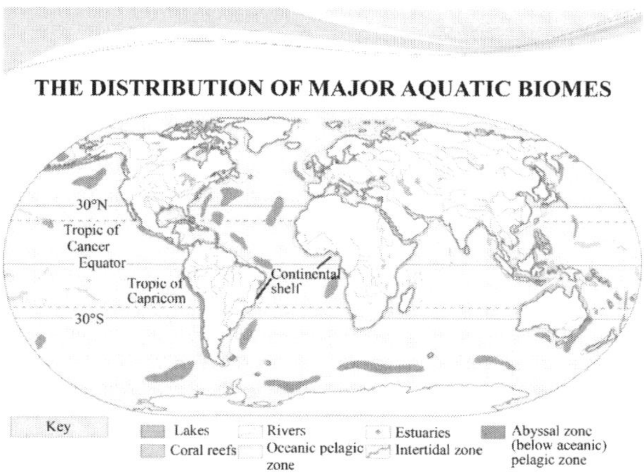

Figure 1.11: Distribution of major aquatic biomes on earth.

https://aquaticmarinebiome.weebly.com/world-map.html)

1.2.2.1 Lakes or lentic biomes include coastal saline lagoons and continental freshwater lakes and lagoons both tropical and templates.

1.2.2.2 Streams and Rivers

These are bodies of flowing water moving in one direction and their characteristics change going from the source to the mouth.

1.2.2.3 Coral Reefs are distributed in warm shallow waters and are found as barriers along continents.

Corals are invertebrates from the Cnidaria occurring as polips that can grow extensive colonies visible from the space. They have been on earth over the past 200–300 million years and in their evolutionary history developed a very interesting symbiosis with Zooxanthellae microalgae. Inside the tissues of each coral polyp live these microscopic, single-celled algae, sharing space, gas exchange and nutrients to survive in a delicate equilibrium.

1.2.2.4 Estuaries and deltas

Estuaries and deltas are areas where freshwater streams or rivers merge with the ocean. There are gradients of salinity from the ocean to the freshwater body. In these areas, different unique communities can be developed including marsh grasses and the topical mangroves.

1.2.2.5 Pelagic Ocean

It is all the ocean that is not close to the bottom nor the coastal areas and constitutes the largest biome on earth. The whales are the biggest mammals found in this biome and most of the ocean fish known as food providers are there. The epipelagic zone is the surficial having a great importance given the phytoplankton and algal communities it harbors and counts as the world largest carbon reservoir by photosynthesis processes.

1.2.2.6 Abyssal Ocean

It is the water strata close to the bottom of the ocean.

1.2.2.7 Intertidal zones

Are also known as the coastal zones affected by the daily movement of the sea level with the tides.

1.3 HISTORICAL BIOGEOGRAPHY

During the 17th century, the job of biogeographers was comparatively a simpler one. The book of Genesis told that "how all men were descended from Noah and that they had made their way from Armenia to their present countries." The consensus was that this was also true for all animals and that they had a common origin from which they too had dispersed as there had been a single geographical and temporal origin for man. After the arrival of 18th century, the impression of the Ark had to be abandoned by people who were informed about natural history. During 1707–1778, a scientist from Sweden named Carl Linnaeus, father of the world biota classification came to rescue the problem. Carl Linnaeus had a lot of faith in God and he believed that God speaks to man through the natural world. In fact, he believed that the universe is a gigantic museum collection, which God has gifted him to define and list into a methodical framework.

Carl Linnaeus gone ahead to solve the Ark problem. By telescoping the story of the Creation into that of the Deluge. He proposed that "all living things had their origin on a high mountain at about the time the primeval waters when they were beginning to recede." As the floodwaters receded, Linnaeus projected many plants as well as animals migrating to their ultimate households where they continued for the rest of the time. For him, "species were fixed entities that stayed just as they were created." Linnaeus also highlighted that each species had been given the structure that was the most appropriate for the habitat in which it lived. To abandon the concept of the Ark, some 18th century naturalists were not ready.

Linneaeus and his students, however, made on of the most extraordinary collections of world living being and classified them. One of his most prolific students was Pehr Loelfing, a Swedish biologist who thoroughly collected the United Kingdom and later Brazil and Venezuela, being the first naturalist in doing collections in one of the oldest regions of the world, today known as the Guiana Shield Ecoregion where he finally died from malaria in the coasts of the Caroni River, a tributary of the Orinoco River.

To understand the pattern of distribution of organisms, it is necessary to be aware that this pattern stems from the interaction of two types of processes, the spatial-temporal processes of living organisms (biotic) and the earth (abiotic) in space over time. However, there are serious difficulties to make the inference of the type of processes that generated the pattern of distribution we see in the present. The greatest difficulty in working with areas of biological science that deal directly with the reconstruction of the past (systematics, biogeography, evolutionary genetics, and evolutionary ecology) is always imposed by the paradox: we could not see the past happening, let alone have some (Fossil shortages), so historical studies are usually based on contemporary information.

There are two explanations for species distribution:

A) Dispersalist explanation, which assumed that there was a pre-geographical area, and there was one ancestral species that evolved in species A and B by way of dispersal to new areas without a geographical barrier. Only after that along time the presence of a geographical barrier between populations A and B has occur speciation. Based on the present evidence, however, we would incorrectly conclude that speciation was conducted by geographic isolation (allopatry, vicariance).

B) Vicariant explanation, there was a pre-geographical area and there was an ancestral species that evolved in species A and B due to the emergence of a geographical barrier that made the ancestral species differ in two species, one on each side of the barrier. Subsequently, this geographical barrier disappeared. Based on the present evidence, we would incorrectly conclude that there was no vicariant or speciation with a barrier because there are no data to show us that the barrier has existed in the past.

From the above history of biogeography, it is possible to summarize that three historical events have had a print in biogeographic scientific research: 1. classical phylogenetic method developed by Hennig, 2. the development of plate tectonics, and 3. Léon Croizat's relentless criticism against the dispersion model: Therefore, three alternative schools emerged from those events:

- Phylogenetic biogeography
- Biogeography of vicariance
- Panbiogeography

1.3.1 Phylogenetic biogeography

During 1784–1785, Thomas Pennant in his Arctic Zoology observed that "the animals of the Ark must have migrated across Asia and then to the Americas via the Bering Strait region." Thus, Pennant was the first to call attention to a migration route that proved to have a great historical importance. Biogeography underwent a fundamental change during the preceding 100 years as seen from the viewpoint of the mid-18th century. In 1766, a German scientist named Johann Reinhold Forster (1729–1798) immigrated to England and published several small works including a volume entitled "A Catalogue of the Animals of North America."

In 1772, he along with his son Georg accompanied Captain Cook on his second expedition to the South Seas, where he presented a universal opinion of the numerous natural regions and their biota. He described "how the different floras replaced one another as the physical characteristics of the environment changed." Forster compared islands to the mainland and based on his observations he noted that "the number of species in a given area was proportionate to the available physical resources." He also observed the uniform reduction in floral diversity from the geographic equator towards the poles and accredited this phenomenon to the latitudinal change in the surface heating of the earth. He found the tropics to be rich, enchanting, and beautiful—"the area in which nature reached its highest and most diversified expression."

Forster understood in a better way that biotas were living communities' characteristic of certain geographical areas, more than any of his predecessors. After which the concept of natural biotic regions was born. Naturalists tended to specialize in the study of either plants or animals as knowledge of the organic world increased and greater numbers of species became known. There were the early botanists for some reason, who took the greatest interest in biogeography. In 1972, a plant systematicist and the head of the Berlin Botanical Garden, Karl Willdenow, outlined the elements of plant geography. He recognized five principal floras in Europe and, like Forster, he was interested in the effect of temperature on floral diversity. Willdenow envisioned an early stage of many mountains surrounded by a global sea to account for the presence of the various botanical provinces.

A very important person in biogeography was Alexander von Humboldt, the most famous student of Willdenow. In his youth, he was overwhelmed and inclined by his friendship with Georg Forster. Von Humboldt felt that "the study of geographical distribution was scientific inquiry of the highest order and that it could lead to the disclosure of fundamental natural laws." He devoted much of his attention to the tropics of the New World following the findings of Loefling in South America, already published by Linneous, and became one of the most famous explorer-naturalists. Von Humboldt included his Essai sur la Geographie des Plantes (1805) as a part of his great 24-volume work Voyage Aux Regions Equinoxiales du Noveau Continent (1805–1837, with A. J. A. Bonpland).

A Swiss botanist named Augustin de Candolle (1778–1841) made the next significant step in the progress of biogeography. Augustin de Candolle in 1820, published his important Essai Elementaire de Geographie Botanique. In that work, he made a distinction between "stations" (habitats) and "habitations" (the major botanical provinces). The botanists Carlos Linneo and Carl Ludwig Willdenow were those who began to group the species of the planet according to their distribution areas in the late 18th century, and in 1855, De Candolle in his work Geography Botany Reasoned, made an interesting model on the genesis of areas of distribution based on species indicative of diverse elements and ecological groups.

In the meantime, the zoologist Georges Cuvier had begun his work on "fossil vertebrates," which was followed by many others later. Cuvier was the world's most influential zoologist in the first half of the 19th century and also a great paleontologist. He was a determined advocate of the Linnaean doctrine of the fixity of species. Cuvier conceived a series of great catastrophes to explain species distribution, the most recent being the biblical Deluge, to account for the discovery of fossils representing unknown species. The new species that later appeared purportedly came from parts of the world previously unknown. Cuvier's catastrophes comprised several regional extinctions. In 1823, was published the work on Reliquiae Diluvianae from William Buckland, who encouraged universal catastrophes as the causes of breaks in the fossil record. The implicit scientific postulate was that "each catastrophe needed a new creation."

In 1813, James Cowles Prichard started to publish his Researches into the Physical History of Mankind. But his most widespread biogeographical conversation was in the second edition of 1826. Prichard was also a catastrophist, but he had interesting explanations for two difficult biogeographic problems already identified: the distribution of island species and the occurrence of disjunct species. These were interestingly explained by Prichard, a catastrophist.

As remote islands support endemic species, they often create problems which appeared to be the result of special creations. According to Prichard observations, "endemic species were in reality only relict populations of species that once had a more extensive distribution." He also felt that as far as plants were concerned, disjunct distributions did not have to be explained by acts of special creation. He pointed out that seeds could be transported by rivers, winds, ocean currents, and also by animals. And he noted that the seeds of some plants could be transported in the digestive tracts of birds.

Charles Lyell (1797–1875) the father of geology and vertebrate paleontology mentioned the relationship between fossil and modern designs from a distributional standpoint. He undertook extensive discussions on botanical geography in his Principals of Geology (1830–1832 and subsequent editions), including the provinces of marine algae, as well as on the geographical distribution of animals. In addition, he examined the effects of geological changes as well as changes in climate on the distribution of species and the sign for the creation and extinction of species. As Browne (1983) has pointed out, the most important was Lyell's suggestion that "the elevation and submersion of large land masses resulted in the conversion of equable climates into extreme ones, and vice versa, according to the quantity of land left above sea level." This understanding meant that floras and faunas had to be active objects, which can contract or expand their borders as geological agents altered climates and topography. So, Lyell brought to biogeography a sense of history and the realization that floral and faunal provinces had almost certainly been altered through time; Lyell was also famous as the champion of gradual change to the earth's surface. As Kinch (1980) has noted and stated, "many early 19th century naturalists expressed in their work the belief that the universe displayed evidence of God's design."

The kind of design that was displayed in nature was not however always agreed by them. One kind has been called utilitarian or purposeful. To explain the distribution of life, the utilitarian group made use of supernatural interventions. In order to prevent the Old World from becoming overpopulated, Thomas Pennant felt that the migration of animals from Asia to America was divinely directed. A surgeon-naturalist aboard the H. M. S. Sulphur, Richard Brinsley Hinds, in 1836–1846, thought "the animals spread from one center, to the resting places of the Ark." The plants simultaneously covered the land surface of the globe. The final statement was based on the biblical statement that "the earth brought forth vegetation."

Two well-known botanists who have searched for an idealist design, which could be discerned as orderly patterns in nature, were Alexander von Humboldt and Robert Brown. They believed that certain plant groups occurred in a given ratio to others and that this ratio would vary in a predictable manner from one geographical area to the next. A publication of Kinch in 1980 showed that during the period 1819–1921, the Quinary system, another idealist design was introduced by William Sharp MacLeary. William Swainson used Quinary system in his biogeographic work. He separated the earth into five major biogeographic regions, which overlapped with his view of the five regions of mankind. In 1832, Edward Newman, an entomologist, advocated a septenary system, but this and the quinary system did not last beyond the arrival of Darwinism.

Edward Forbes (1815–1854) made significant aids to both marine and terrestrial biogeography, despite of his short life. He accounted for the clear connection between the European mountain tops and Scandinavian floras by supposing land subsidence and the cold conditions in the fresh past. His work on "mapping of the distribution of marine life together with a descriptive text" appeared in Alexander K. Johnston's "The Physical Atlas of Natural Phenomena in 1856," which included the foremost complete work on marine biogeography. In his work, the world was divided into 25 provinces and was positioned within a series of nine flat "homoizoic belts." A five depth zone series was also documented. In the same year, the famous malacologist, Samuel P. Woodward published his third edition of "Manual of the Mollusca," which included with the universal circulation of the Molluscan group.

In 1859, Forbes delayed work was published by Robert Godwin–Austen, the posthumous work on "The Natural History of European Seas." Forbes has made several observations on his work, these observations are:

- Each zoogeographic province is an area where there was a special manifestation of creative power and that the animals originally formed there were apt to become mixed with emigrants from other provinces;
- Each species was created only once and that individuals tended to migrate outward from their center of origin; and
- Provinces, to be understood, must be traced back like species to their origins in past time.

During 1838–1842, another imperative contribution was made by James D. Dana who participated in the US Exploring Expedition. He was able to divide the surface waters of the world into several different zones based on temperature and used isocrymes (lines of mean minimum temperature), through observations made on the distribution of corals and crustaceans. In 1853, his brief paper related to his plan was published in the American Journal of Science. Ludwig K. Schmarda in his volume entitled Die Geographische Verbreitung der Tierre in1885 included all data related to animal life, marine, and terrestrial in a single zoogeographic scheme. He divided the world into 10 marine realms and 21 terrestrial realms.

The terrestrial world was then divided into the biogeographic regions that, essentially, are still in use today was divided by P. L. Sclater. This was completed in 1858 in a small paper entitled "On the General Geographical Distribution of the Members of the Class Aves." Sclater's work proved to be useful for almost all groups of terrestrial animals, despite the fact his scheme was based only on the distributional patterns of birds. This distribution of animals in different biogeographic boundaries has helped very much scientists to find the animals in their respective locations.

During 1835, when the young Charles Darwin visited the Galapagos Islands, he was amazed by the individuality, yet basic resemblance, of the fauna to that of mainland South America. Darwin was among the first scientists who certainly appreciated the importance of a geographic scale. In the support of evolution, this can be explained appropriately by taking the example of "Charles Darwin's study of finches" on the Galapagos Islands, which resulted in his book *On the Origin of Species*. As noted by Darwin, the finches on the mainland of South America were similar to those located on Galapagos Islands, geographically and morphologically distinct, however, depending on the type of food available on each island, their shape of the beaks differ. Initially, the islands were part of South American mainland, but, subsequently, two landmasses were separated and drifted apart. This resulted in creation of novel habitats and food sources obtainable for the species present in each of these regions. Therefore, by the genetic processes and through the selection of alleles, which promoted survival each finch species had adapted to the local environment, ultimately resulting in speciation.

To study biogeography, islands are excellent, as they comprise of ecosystems that can easily be linked to those of the mainland as well as the other surrounding regions. Early biogeographers who recognized certain types of distributions of organisms have made these sorts of observations. There were some species, which were restricted to a particular region only, and are found in that region only and nowhere else, these species were referred to as endemic species. With relation to the taxonomic group, the term endemism needs to be defined that: "all life forms that are known are endemic to the planet earth; the genus *Geospiza* (Darwin's finches) are constrained to the Galapagos islands; *Geospiza fortis* is endemic to those specific islands; the spotted owl is endemic to the old-growth forests of the pacific northwest." The distribution of cosmopolitan species is universal. Cosmopolitan species generally occur in most continents but are restricted to specific habitats only.

Another important pattern that needs to be defined here, in addition to endemism, is a disjunct distribution; this term defines clearly related species or even the same species that are found in diverse areas. For example, Ratite birds (Ostrich, Emu, Cassowary, Rhea) are found in Africa, Australia, and South America, respectively, and Marsupials are found in Australia and South America.

Some 20 years later after Darwin's work, Alfred Russel Wallace while traveling through the Indo–Australian Archipelago, he was puzzled by the contrasting character of the island faunas, some with Southeast Asian affinities and others with Australian relationships. As noticed by Alfred

Russell Wallace, "different regions of the world had consistent patterns of endemic species" and based on his observations he drew up six biogeographic realms. Primarily, Wallace worked in Malaysia, where he noticed a clear break between the fauna on the islands to the northwest and the Australian fauna. This break has later come to be identified as Wallace's line (also a line between the Asian and the Australian biogeographic zones). These patterns were defined long before continental drift was a problem.

For levels of similarity in the biota, different biogeographic areas can be quantified. In 1848, the evolution was already accepted by Alfred Russell Wallace when he began his travel through the Southeast Asia and Amazon. On his journeys, he wanted to prove that evolution certainly takes place, by showing how geography affected the different range of species. He studied hundreds and thousands of plants and animals, carefully noticing accurately where he had found them. The patterns founded by him were convincing evidence for evolution. He was struck, for example, by how rivers and mountain ranges marked the boundaries of many species' ranges. As he could find similar climatic regions with very different animals in them, the conventional explanation that species had been created with adaptations to their particular climate made no sense. Wallace came up with a similar conclusion as detailed by Darwin in his publication the *Origin of Species*, the conclusion was: "Biogeography was simply a record of inheritance, as species colonized new habitats and their old ranges were divided by mountain ranges or other barriers, they took on the distributions they have today."

After careful consideration on their work, to account for evolutionary change, both Darwin and Wallace arrived at a theoretical mechanism, which is also known as natural selection. The key for both Darwin and Wallace was "the realization that distributional patterns had evolutionary significance." In 1858, Darwin and Wallace have announced the work on their joint theory in the Journal of the Linnean Society of London and, especially in the publication of "Darwin's Origin of Species" in 1859, which changed the thinking of the civilized world. The Darwin's Origin of Species book comprises two important chapters on geographical distribution. Darwin made three important points in discussing biogeography from the viewpoint of evolutionary change, the three key observations are:

- He highlighted that barriers to migration allowed time for the slow process of modification through natural selection;
- He considered the concept of single centers of creation to be

critical, that is, each species was first formed in one area only and they migrate from that center to as far as their ability would allow;

- He noted that dispersion was a phenomenon for global rank.

In regard to the third point, "Darwin observed that oceanic islands were generally volcanic in origin and must have accumulated their biota by dispersal from some mainland source."

The closest confidants of Darwin while he was going through the prolonged process of formulating his theory were Joseph D. Hooker and Charles Lyell. Joseph D. Hooker, a great plant collector and systematist, was particularly interested in southern hemisphere botany. Hooker also accompanied Sir James Ross on his Antarctic Expedition (1839–1843). However Hooker felt that Darwin was too reliant on dispersion in accounting for disjunct relationships. Hooker speculated on the likelihood that the Southern Ocean plants were the remnants of a flora that had once been spread over a more continuous and a larger tract of land than now happens in that part of the world while describing the flora of New Zealand in 1853. In contemporary terms, he was signifying a vicariant rather than a dispersal history for the sub Antarctic floras.

Alfred Russel Wallace applied himself primarily to biogeography whereas Darwin also applied on biogeography going to investigate many other aspects of evolutionary change. In the year 1876, Wallace finally published his epic two volume work "The Geographical Distribution of Animals" where he worked to reach a number of assumptions about biogeography that are still worth reviewing. For example, he pointed out that:

- It is important to have the idea about how paleoclimatic studies are very important for analyzing extant distribution patterns;
- For distribution, dispersal, and extinction of animals and plants, factors like competition, predation, and other biotic factors play important roles;
- Discontinuous ranges may come about by extinction in intermediate areas or patchiness of habitats;
- Dislocations of genera show superior antiquity than those of a single species, and so forth for advanced groupings;
- The joint occurrence of organisms not adapted for long distance dispersal is good evidence of past land connections;

- Extinction may occur, when two large land masses long separated are reunited, as many organisms will come across new competitors;
- Islands may be classified into three major categories: "continental islands recently set off from the mainland," "continental islands long separated from the mainland," and "oceanic islands of volcanic and coralline origin";
- Because the relationships among distribution, speciation, and adaptation are easier to see and comprehend, studies of island biotas are important.

Alfred Russel Wallace, in his work presents a map in Mercator projection with the zoogeographic zones of the planet. This map shows a simple biogeographic scheme in which Earth is divided into six kingdoms. The American continent is divided into two kingdoms separated by the Tropic of Cancer, to the north the Neartic region and to the south to the Neotropical region. Europe, much of Asia and the territories located above the tropics in Africa make up the Paleoartic region. Most of Africa along the island of Madagascar forms the region of Ethiopia. India, southern China and the Philippines make up the eastern region and finally the islands of northern Australia, the Australian continent itself, New Zealand and the island of Tasmania are part of the Australian region.

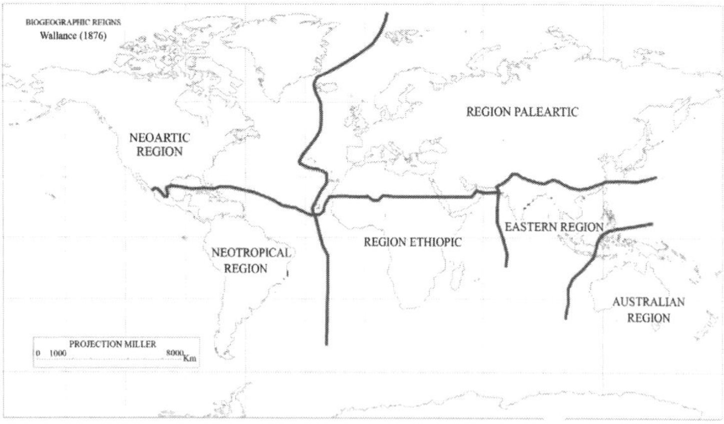

Figure 1.12: The biogeographical reigns of Alfred Russel Wallace.

(adapted from http://www.geo.arizona.edu/Antevs/ecol438/walcregn.html)

Another work was done by Ernst Haeckel in 1876 where he published his curious work on "The History of Creation," or "The Development of the Earth and its Inhabitants by the Action of Natural Causes.." In Lemuria, a continent in the center of the Indian Ocean, Ernst Haeckel placed the location of Paradise, where life was created. He offered a map showing "how humans and the rest of the biota reached other parts of the world." In 1907, he relocated Paradise to the north and east of India in a later edition of his book. In 1927, Hermann von Ihring resurrected the Lemuria concept. Many biogeographers repeated Wallace distribution plan without any major new interpretations as well as have been following the publication of Wallace's works. In 1890, E. L. Trouessart published his "La Geographie Zoologique" which inspected both marine as well as terrestrial patterns. In 1895, Frank E. Beddard came out with "A Textbook of Zoogeography." In 1907, Angelo Heilprin published a volume entitled "The Geographical and Geological Distribution of Animals." Some minor changes were introduced by the latter in Wallace's map and also studied the data then available about the distribution of fossil forms. Also, a number of works have been published, dealing with the rise and fall of mid-ocean continents and the establishment of hypothetical land bridges. These theories were discarded as the knowledge of sea-floor history increased.

Building on the foundation provided by Wallace and Darwin, plant geography also made significant progress. As shown by the work of A. Grisebach in 1872, Die Vegetation der Erde nach ihrer Klimatischen Anordnung, the 1879–1882 works of A. Engler, Versuch einer Enwicklungsgeschichte der Pflanzenwelt, and the 1890 book by O. Drude, Handbuch der Pflanzengeographie, and the German school was particularly active. Numerous phylogenetic trees demonstrated the biological relationships amongst the landforms. Arldt assumed a history of stable continents that were connected by land bridges, in order to account for such relationships. Later, Hermann von Ihring converted these land bridges into imposing mid-oceanic continents in his 1927 work "Die Geschichte des Atlantischen Ozeans." He outlined an Archhelenis in the South Atlantic, an Archatlantis in the North Atlantic, and a Lemuria in the western Indian Ocean. Supposedly, these landforms occurred from the Cretaceous to the Eocene.

In 1915 when William Diller Matthew (1871–1930), a geologist and paleontologist, published his article on "Climate and Evolution," a significant advance in biogeography took place. Matthew was an expert on fossil mammals and his 1915 work was dedicated principally to highlighting the rank of the northern hemisphere in the evolution as well as dispersion of that group. Yet, the most significant feature of that work has turned out to be Matthew's declaration of his concept about centers of dispersal. He said, "At any given period, the most advanced and progressive species of the race will be those inhabiting that region; the most primitive and unprogressive species will be those remote from this center. The remoteness is, of course, not a matter of geographic distance but of inaccessibility to species invasion, conditioned by the habitat and facilities for migration and dispersal."

Also in 1915, the "age and area" hypothesis was developed by botanist J. C. Willis. In 1922, the book on "Age and Area: A Study in Geographical Distribution and Origin of Species" was published. In this hypothesis was stated according to which the areas engaged by a given group of related species depends upon the age of the species in that particular group; that is, "the older the species, the greater the occupied area." In the early stages, this theory gained lot of attention but later it was realized that many species occupying partial areas were geographical relicts. The outcome was that two kinds of species were found which were confined to smaller areas, these were, the old ones that were formerly dispersed but had become extinct in most parts of their range and the newly evolved ones that had not yet become widespread. Arnold Ortman made the progress in the knowledge about distribution patterns in the marine environment, as he published his "Grundzuge der Marinen Tiergeographie (1896)." The subsequent year, that is in 1897, a paper on the distribution of marine mammals by Philip L. Sclater. During the year 1935, Sven Ekman accomplished the huge task of studying all of the appropriate literature on marine animal distribution and published the results in his book entitled "Tiergeographie des Meeres," a second edition of which was printed in the year 1953 in English. Researchers like John C. Briggs, Marine Zoogeography (1974), Geerat J. Vermeij, Biogeography and Adaptation (1978), S. van der Spoel and A. C. Pierrot-Bults (eds.), Zoogeography and Diversity in Plankton (1979), Oleg G. Kussakin (ed.), Marine Biogeography (1982, in Russian), and A. C. Pierrot-Bults, et al. (eds.), Pelagic Biogeography (1986) have published the modern books on marine zoogeography.

A new progress took place which united the quickly developing field of ecology with biogeography in the year 1920s and 1930s. In 1921, Friedrich Dahl's Grundlagen einer okologischen Tiergeographie, and in 1924, Richard Hesse's Tiergeographie auf okologischer Grundlage marked the beginning of biogeography. These efforts were in reply to a need to inspect the geographical distribution of animal and plant community on a local as well as on a universal scale.

In 1950, almost one hundred years after Darwin published The Origin of Species, the German entomologist Willi Hennig proposed a classification method to reconstruct the history of life, even if only the data of the present could be seen from it. Systematics, a discipline that attempts to classify living beings according to criteria that vary with the epochs, is qualified, in the present case, as phylogenetic. The method proposed by Hennig is based on the fact that, due to evolution, each biological structure is defined, in a historical sense, according to the place it occupies in the family tree of living beings. The unique biological structures and functions that inherited from the ancestral species allow the characterization of these ancestral species as those that descend from them. Recognizing such unique characteristics, it is possible to detect the genealogical or phylogenetic relationships among species.

In 1967 Georges Lemée, representing the French school published the book Précis de biogéographie, in which he divides the earth in five faunistic and floral empires, some of them separated with several transition zones along the deserts of Sahara, Arab and Gobi and the islands of Indonesia and Melanesia. The classification of Lemée includes both geobotanic and zoogeographic criteria and the author incorporate the elements of endemicity.

1.3.2 Biogeography of Vicarianism

León Croizat was the father of the vicariant biogeography. This eccentric botanist and biogeographer laid the groundwork for what we understand today as Vicariant biogeography. He said one of the most famous phrases ever delivered in Biogeography: "life and earth evolve together."(Croizat, 1964). This Italian, eradicated in Venezuela where he lived to the end of his life, wanted to show the world that life and the planet are correlated in their history. That is, the geological history can help us understand the history of organisms, as well as the history of organisms can help us to understand the history of our planet. Even before the knowledge of continental drift and plate tectonics, Croizat already showed, by his studies of pattern of distribution of plants, the Continental Drift or "how the continents moved along earth history."

Croizat has come to the conclusion that Oceans Atlantic, Indic and Pacific were closer to each other in the past than in the way they are known today. From the precepts of Croizat arises the vicarious thought, opposed directly to the Darwin's dispersalist thinking, that was the dominant paradigm at the time.

Through a combination of historical factors like speciation, extinction, continental drift, and glaciation the patterns of species distribution across geographical areas is usually explained. Associated variations in sea level, river routes, habitat, and river capture can be seen by observing the geographic distribution of species. Over stages of ecological changes, "biogeography includes the study of plant and animal species in their past and/or present living refugium habitat; their interim living sites; and/or their survival locales." To understand the factors affecting organism distribution, and to predict future trends in organism distribution, modern biogeography frequently employs the use of Geographic Information Systems (GIS). Also, to solve ecological problems that have a spatial aspect to them, mathematical models and GIS are employed. More recently, neutral models have been used to test important mechanisms driving diversity that are causing new revolutions and re-thinking ideas about explanations of patterns in biogeography and biodiversity.

Biogeography is often an imperative part of analysis for scientists who study the most important and advanced theory in ecological knowledge, the Theory of evolution. Biogeography provides convincing proofs for that theory. This is because many geographical features like oceans, rivers, mountains and islands provide barriers to species, letting scientists to detect how they evolve from one another.

By studying the island and continental biogeography come the most significant pieces of biogeographical proof for evolution. Darwin's important discoveries have occurred on remote islands, like Galapagos. Darwin noticed that there were unique species in these remote locations, which were not found in any other location even in the similar climate zones elsewhere on Earth. It is from this insight that much of evolution's most important biogeographical proof emanates. Darwin sought to answer the research question, "Why do animals on distant and isolated landmasses appear related, but distinct?" Evolution as a biological process was his answer.

Charles Darwin used isolated oceanic islands since the beginnings of the theory of evolution, most importantly "South America's Galapagos Islands" to show how new species were evolved from isolated environments. Darwin asked the question in the case of the Galapagos that "why the Galapagos and the Cape Verde Islands, which are off the coast of northwestern Africa, have such distinct species, despite having nearly identical climates." More prominently, Darwin saw that the species on both islands seemed to be closely related to the species on the nearest continent. Darwin concluded that the animals on these isolated islands must have been originally from the nearby continent, but they gradually evolved into something different over thousands of years because they were separated from the other species on the continent.

Another famous research question was "how an isolated region seems to produce different animals that are nonetheless clearly related to the nearest larger landmass" and the answer to this were the marsupials of Australia. It is clear that despite being thousands of miles away marsupials in South America and Australia appear to be related though the exact lineage of marsupials is still being debated. This relatedness is probably due to plate tectonics which was hard to understand by Darwin during that time. When South America and Australia were amalgamated in a single continent, an "original" marsupial lived there, and when they separated, the marsupials on each continent gradually evolved into distinct species and by the evolutionary process they got better adapt to their new environments as the two continents separated.

For Darwin, the fact that mammals, except when they were introduced by humans, were almost never naturally present on islands that were more than 300 miles from the nearest landmass is one of the most significant pieces of biogeographical evidence in favor of evolution. To answer another important research question: "Why were there no mammals on islands like the Canary Islands or the Galapagos?" For Darwin, the answer was simple since it needed traveling over large masses of water; it was hard for large terrestrial animals to get to isolated islands that were 300 miles from land. As such, the lack of mammals on islands suggests that mammals never made it to such islands, and thus never had a chance to evolve and adapt to those environments.

The enthusiastic promotion of the theory of vicarianism was the most important and controversial development of the decade of the 1970s in the 20th century. "Vicariance refers to the biogeographic patterns produced by a particular kind of allopatric speciation in which a geographic barrier develops so that it separates a formerly continuous population." This differentiates vicarianism from the kind of dispersalism, which takes place as the consequence of dispersal of individuals across an existing barrier to colonize the other side or by migration. Although these two kinds of allopatric speciation had been known for several years, the supporters of vicarianism, came to feel that the dispersalism viewpoint had been deserted and the vicarianism was the imperative method in reconstructing evolutionary change.

León Croizat was an eccentric and controverted botanist and biogeographer laid the groundwork for what we understand today as Vicariant biogeography. He said one of the most famous phrases ever delivered in Biogeography: "life and earth evolve together."(Croizat, 1964). This Italian, from Turin, although his father wanted him to be an engineer, by destiny he became a botanist, abandoned his country of origin and went to United States where he worked at the New York Botanical Garden and had the possibility to analyze the botanical collections, identify plants and write many of his publications. He was trying to return to his country but still there were bad conditions from the war and he asked Prof. Henry Pittier to work as botanist at the Botanical Garden of Venezuela in Caracas and eradicated in Venezuela where he lived to the end of his life and was the inspiration for many botanist and zoologists of the late 20th and the beginning of the 21st centuries. In particular his work in Venezuela excited the development of many hypotheses to explain the high endemism and paleogeographical evolution of birds and plants biota communities of Pantepui, a group of table mountains up to 2.700 m a.s.l. separated among them as islands in a vast matrix of forested landscapes in the world oldest Precambrian Guiana Shield. He wanted to show the world that life and the planet are correlated in their history. That is, the geological history can help us understand the history of organisms, as well as the history of organisms can help us to understand the history of our planet. Even before the knowledge of continental drift and plate tectonics, Croizat already showed, by his studies of pattern of distribution of plants, the Continental Drift "how the continents moved along earth history."

By envisioning, a connection between plate tectonics and vicarianism was recognized, before the separation of the continents, a hologenesis: "Hologenesis is referred to as a place where species were supposed to have been created with cosmopolitan ranges, may be contrasted with the center of origin concept where species originated in a limited area and then spread as far as their capabilities would permit" as espoused by Rosa (1923). The supporters of vicarianism emphasized that it was really vicariance that produced multiplication of species and geographical differentiation while dispersal produced only sympatry, as their enthusiasm for a supposedly new concept grew. In 1974, Croizat, Nelson, and Rosen in the best explanation of the mechanics of vicarianism, stated that, "The existence of races or subspecies that are separated by barriers (vicariance) means that a population has been subdivided, or is subdividing, not that dispersal has occurred, or is occurring across the barriers."

Trust in vicariance led its followers to uphold that centers of origin do not occur since, to identify such centers, they would have to acknowledge that species are proficient of scattering from their places of origin to create themselves elsewhere, the usual result being, after a period of time, allopatric speciation by migration rather than by geologic change. Accordingly, Croizat et al. said, "We reject the Darwinian concept of the center of origin and its corollary, dispersal of species, as a conceptual model of general applicability in historical biogeography." Many journal articles were published in the late 1970s and 1980s about the pros and cons of vicarianism.

By the 1970s, this work began to have a noteworthy impression on the practice employed by people who did systematic work. A set of rules for the practice of systematics have been provided by Hennig which have collectively been called "cladism." These rules have been helpful but the one that applies to biogeography has turned out to be suspect. It states that "species possessing the most primitive characters are found within the earliest occupied part of the area, that is, the center of origin for that group." Although at first, this rule was actively accepted by some, a very little biogeographical evidence has been found to support it. A modification in the attitudes of the principal vicarianists became apparent in the 1980s. "Vicarianists now admit that allopatric speciation via dispersal can also take place and only maintain that it is less important than vicariance" as said by McCoy and Heck (1983). Cracraft (1983) specified that to join forces with the vicarianists, some cladists were able to sacrifice their center of origin concept. But in his book about biogeography and plate tectonics, Briggs emphasizes how Humphries and Parenti considered dispersal biogeography to be an unscientific, ad hoc discipline that "… can never let us discover the history of the earth." Vicariance hypotheses were considered as scientific because it was testable, and two tests may be applied to a vicariance hypothesis: "add more tracks (reinforcement by other taxa that show the same pattern)" and "compare the hypothesis to a geological one." It must here be highlighted that any biogeographic hypothesis is on shaky ground if it is based only on the distribution and relationships of a single group of organisms. The strongest hypotheses are those based on common patterns proven by many diverse biotic groups and are, at the same time, consistent with a well-substantiated geological history.

Vicariance biogeography has typically converted into cladistic biogeography. The latter takes phylogenies, built according to the guidelines of Hennig, and combine them with the vicariance model. There are two general propositions of vicariance model or paradigm embodies:

- The best first-order explanation for an observed disjunct distribution among sister groups is that it represents a fragmentation of a widespread ancestral species, rather than a dispersal phenomenon from a more restricted center of origin;
- The more cases of such observed patterns, the more likely the first proposition becomes. The difficulty with the modern vicariance model is that it asks the investigator to perform an act of faith before he begins his analysis; he must believe in a given first-order explanation.

1.3.3 Biogeography of Panbiogeography

The objectivity of the vicariance, however, is questionable. "Suppose a researcher has analyzed the relationships within a given taxon so that he has been able to construct a phylogenetic tree or a cladogram. If the work is to have a biogeographic value, the geographic patterns of the species need to be related to the evolutionary ones. How should this be done?" An objective approach is to sacrifice a priori expectations and let the group being worked on, suggest what the history might have been with comparisons to similar groups. The resulting hypothesis will involve both vicariance and dispersal, if the group is an old one, and if appropriate consideration has been paid to interrupting factors (paleontological, geological, climatological). This uncommitted approach is named as "eclectic biogeography." Simpson by 1980 observed that a reasonable biogeographer should be an eclecticist, which means neither a vicarianist nor a dispersalist." The vicariance model, places two significant limitations on the expansion of biogeography as a field of research:

- First, "it recognizes only a certain kind of allopatric speciation as being of historical importance."
- Second, "it does not recognize that centers of origin exist since, to do so, would admit that dispersal has taken place."

The natural biogeographic regions of the globe are represented by ocean basins as believed by supporters of panbiogeography.. Also the land masses are biological and geological composites located at the boundaries of the basins.

1.4 PRESENT WORK: AN EVOLUTION BIOGEOGRAPHY, THE LINK BETWEEN BIOGEOGRAPHY AND BIODIVERSITY

Molecular analysis, Cladistics, and Panbiogeography are the methods for biogeographers to study distributional patterns alone without any particular connection to phylogenetic patterns.

Current patterns of distribution of the species can be explained by processes of the earth that occurred along its history. Plate tectonics dynamics being the higher driver for these processes as continental drift created bridges and barriers. Also, the climatic changes of the Cenozoic (glacial and interglacial periods of the Tertiary and Quaternary,drove the current diversification and extinction of the groups used in biogeographic classifications (mammals and angiosperms).

The first large period of the earth was the Permian during which the continents were all together being part of what it has been called as the Big Continent Pangaea.

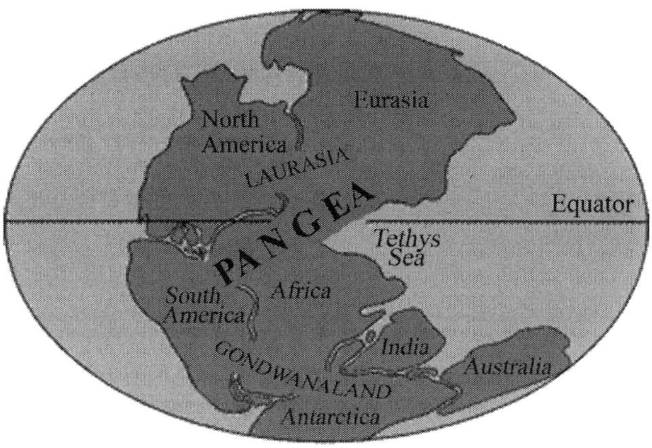

Figure 1.13: Pangea Supercontinent.

(Source: https://www.isgs.illinois.edu/outreach/geology-resources/plate-tectonics-mysteries-solved)

Table 1.1: The history of earth and major events

https://www.isgs.illinois.edu/outreach/geology-resources/plate-tectonics-mys-teries-solved

		Millions of years before present day (BP)		
Precambric		4.500–570		
Phanerozoic	Cambric	570–505		
	Ordovicic	505–438		
	Siluric	438–408		
	Devonic	408–360		
	Carboniferous	360–290	Rainforest collapse Reptils diversification with the evolution of Dinosaurs	
	Permic	290–250		
	Triassic	250–205		
	Jurassic	205–135		
	Cretacic	135–66	Cretaceous Terrestrial Revolution	Dramatic changes in Marine Realms
	Tertiary	66–24		
	Quaternary	2–0.01		
	Holocene	0.01–Present		

During the Triassic Period started the movements of the tectonic plates and continents major divisions: with it all animals and plant species that were present were divided by major ridges and/or isolated in islands. It was the time of Laurasia and Gondwanaland.

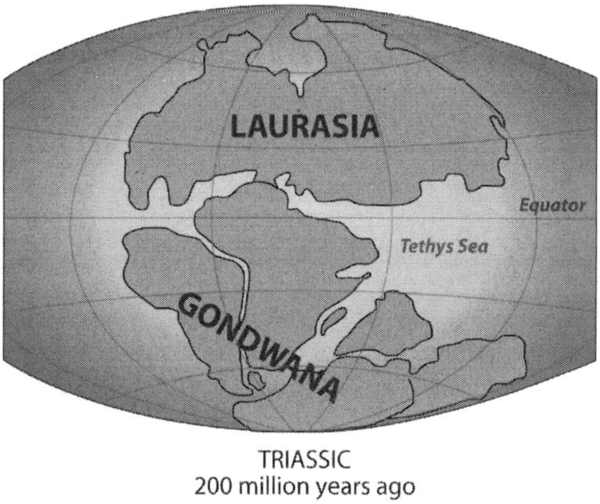

Figure 1.14: The distribution of continents during the Triassic Period.

(Source: http://cdn.zmescience.com/wp-content/uploads/2013/09/laurasia-gondwana.jpg)

During the Cretaceous occurred the evolution of higher flower plants and mammals and there was the rise of the Atlantic Ridge that produced the split of the tectonic plates of Africa and South America creating the Andes Mountains when the Plate of South America crashed against the Plate of the Pacifics. It was also the time seeing the last movement of India to North until it was unified with Laurasia creating the Himalayan Mountains.

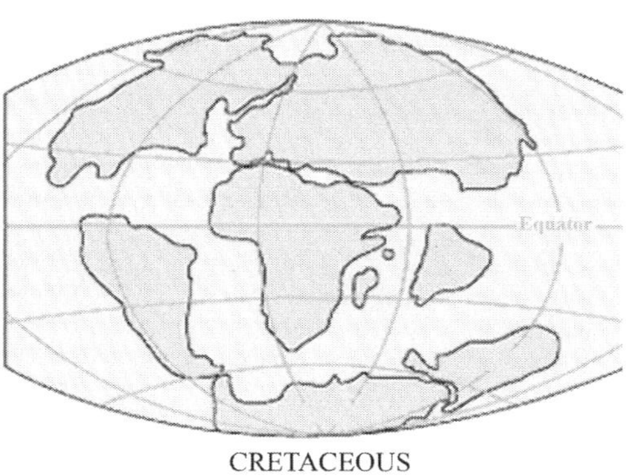

Figure 1.15: The distribution of continents during the Cretaceous Period.

(Source: http://www.marinebio.net/marinescience/02ocean/mgimg/pancre4.jpg)

Every continent has its distinct latitudinal variation, rivers, lagoons and coasts, topography with mountains and altitudinal belts as a result of which different kinds of vegetation and animals abound in each region. For instance, diversification of mammals occurred during the Early Cenozoic Period (50–60 millions of years BP) as it can be seen in Figure 3.15.: that is the reason why in Africa we have gorillas, lions, zebras and rhinoceroses and these are nowhere to be seen in South of America. Instead, there are tapirs, capybaras, manatees and jaguars there.

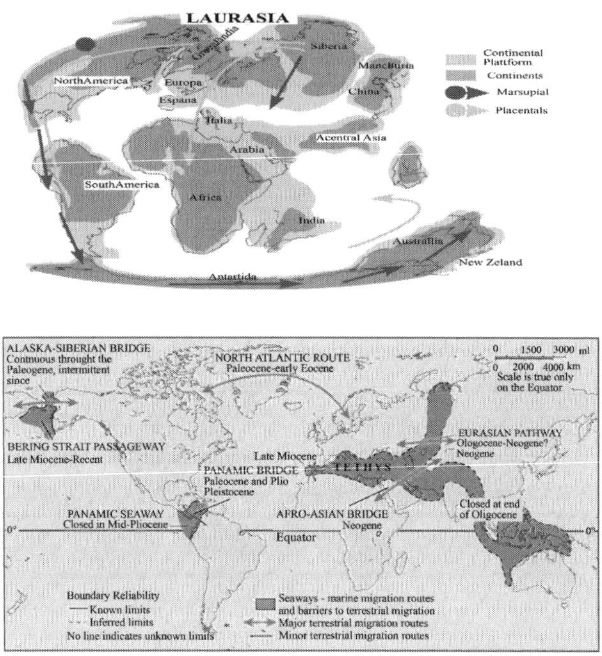

Figure 1.16: Distribution and routes of migration of Mammals during the Cenozoic.

(Source: https://media1.britannica.com/eb-media/13/813–004–10B408EB.jpg)

C. Barry Cox and Peter D. Moore write in their text *Biogeography: An Ecological and Evolutionary Approach*, 7th edition: "The patterns of biogeography are the result of the interaction between the two great engines of our planet: evolution and plate tectonics.... Because it faces such wide-ranging questions, biogeography must draw upon an extensive range of other disciplines. Explaining biodiversity, for example, involves the understanding of climate patterns over the face of the earth, and the way in which the productivity of photosynthetic plants differs with climate and latitude. We must also understand what makes particular habitats desirable to biota; why locations of particular soil chemistry, or moisture levels, or temperature range, or spatial structure, should be especially attractive. Hence, apart from earth and geographical, also biological and environmental sciences must be invoked to answer such questions.

Biogeography, then, is concerned with the analysis and explanation of patterns of distribution, and with the understanding of changes in distribution that have taken place in the past and arc taking place today."

In this 21st century, Biogeography embraces several diverse arenas including geology, systematics, botany and plant biology, zoology, and general biology. "A biogeographer's main focus lies on the large-scale biological variation, explaining distribution patterns in terms of which environmental factors and how the influence of humans does to the distribution of the species." Technological advances from the last decade of 20th century up to the second decade of 21st have allowed to use satellite imaging and processing of the earth, in terms of applications of biogeography as a science today. Global Production Efficiency Model (GLO-PEM) and General Information Sensing (GIS) are very important within modern biogeography. GLO-PEM uses satellite-imaging giving "repetitive, spatially contiguous, and time specific observations of vegetation." These observations are on a global scale: it has been demonstrated for instance that whale locations relate to sea surface temperatures, when processes on the earth's surface are temporally analyzed with GIS horizontal using digital images spatial distribution and vertical depths by bathymetry.

Biogeography addresses many important topics at a time when human culture is increasingly capable of altering the atmosphere, hydrosphere, and lithosphere that form the underpinnings to the pattern of life on the planet; this is why biogeography is considered as a growing and maturing field of study. The revolutionary work in biogeography was carried out at a time when sailboats accomplished travels by foot, as well as horses, buggies. All of that has changed because of technology; today people use jet aircraft and automobile to travel around the world. Because of satellite imagery, our culture has a new perspective of the planet. In a matter of decades, technology has changed the quality and content of what was being perceived from nature as well as the way in which the world was being perceived. Although in the recent past biogeographers focused their efforts on the natural patterns of communities of undisturbed organisms, in contemporary times no place is untouched by human activities. Most of the communities are increasing threatened with extinction and some of them have already been disappeared.

Technological development has progressed to the opinion “that the very life support system of the planet is at risk of destruction.” It is imperative that society gains perspective on the life qualities of the planet so that the denotation and significances of the changing world can affect strategies as well as movements by the world community. Biogeography emphasizes consideration on methods which produce the improved assets of the planet that allow designs of life to thrive. “The study of biogeography shows us that, throughout the ages, life has existed in a wide variety of changing and impermanent patterns; that patterns of life are now increasingly being determined by human technological development; and that perfect understanding of the final truth about the intricacies of life may never be reached.” The Earth and its life-enhancing qualities are the most valuable things and nothing is more precious than this.

As we will see in the next chapter, the concept of biodiversity refers to the variety of life at different scales of time and space. The most recent and revolutionary work in directly relating biogeography and biodiversity is developed in the Neutral theory of Hubell. This author gives random events explanations to the dynamics of births, mortality, speciation and dispersion in explaining the local structure of ecological communities and the dynamics at the level of what is called meta-communities that are distributed in large geographical scales.

REFERENCES

1. *Biogeography: Wallace and Wegener*. Retrieved October 14, 2017, from http://evolution.berkeley.edu/evolibrary/article/history_16
2. Briggs, J. (1987). *Biogeography and plate tectonics*. Elsevier. Retrieved October 14, 2017, from http://file:///C:/Users/Cognitive/Downloads/J.C.%20Briggs%20Eds.%20Biogeography%20and%20Plate%20Tectonics.pdf
3. Briggs, J. (1995). *Global biogeography*. Retrieved October 14, 2017, from http://file:///C:/Users/Cognitive/Downloads/John%20C.%20Briggs%20Eds.%20Global%20Biogeography.pdf
4. Connell, J. (1978). Diversity in tropical rain forests and coral reefs. *Science* (new series), *199*(4335), 1302-1310.
5. Crisci, J. V. (2001). The voice of historical biogeography. *Journal of Biogeography*, *28,* 157-168.
6. Crisci, J. V., Katinas, L., & Posadas, P. (2003). *Historical biogeography: An introduction*. Cambridge: Harvard University Press.
7. Croizat, L. (1958). *Panbiogeography* (Vols. 1 and 2). Caracas, Venezuela: Author.
8. Croizat, L. (1982). Vicariance/vicariism, panbiogeography, vicariance biogeography, etc.: A clarification. *Systematic Zoology*, *31*, 291-304.
9. Croizat, L., Nelson, G., & Rosen, D. E. (1974). Centers of origin and related concepts. *Systematic Zoology*, *23*, 265-287
10. Darwin, C. R. (1859). *The origin of species*. London: John Murray.
11. Hill, J, & Hill, R. (2001). Why are tropical rain forests so species rich? Classifying, reviewing and evaluating theories. *Progress in Physical Geography 25,* 326-354.
12. Hubbell, S. P. (2001). *The unified neutral theory of biodiversity and biogeography* (p. 392). Princeton: Princeton University Press.
13. Hubbell, S., O'dwyer, J., Okie, J., Ostling, A., Ritchie, M., Storch, D., & West, G. (2014). On theory in ecology. *BioScience*, *64*(8), 701-710.
14. Huber, O. (1987). Consideraciones sobre el concepto de Pantepui. *Pantepui*, *2*, 2-10.
15. Marquet, P., Allen, A., Brown, J., Dunne, J., Enquist, B., Gillooly, B., Gowaty, P., Green, J., Harte, J.,
16. Mayr, E., & Phelps, W. H. (1967). The origin of the bird fauna of the

southern Venezuelan highlands. *Bulletin of the American Museum of Natural History, 136*, 269-328.

17. Nelson, G., & Platnick, N. I. (1981). *Systematics and biogeography: Cladistics and vicariance*. New York: Columbia University Press.
18. Odum, E. P., & Barrett, G. (Illustrator). (2004). *Fundamentals of ecology* (5th ed., p. 624). Belmont, CA: Brooks Cole. ISBN 0534420664.
19. Rosen, D. E. (1978). Vicariant patterns and historical explanation in biogeography. *Systematic Zoology*, *27*, 159-188.
20. Rull, V. (2005a) Biotic diversifications in the Guayana Highlands: a proposal. *Journal of Biogeography,* 32, 921-927.
21. Steyermark, J. A. (1983a). Dr. León Croizat-Chaley. *Natura* (Caracas), *74,* 34-35.
22. Steyermark, J. A. (1986). Speciation and endemism in the flora of the Venezuelan tepuis. In F.
23. Stiling, P. (1995). *Ecology: Theories and applications* (3rd ed., p. 638). New Delhi: Prentice Hall.
24. ThoughtCo. (2017). *Biogeography provides evidence of evolution.* Retrieved October 14, 2017, from https://www.thoughtco.com/how-biogeography-supports-evolution-249883
25. Wallace, A. R. (1876). *The geographical distribution of animals: With a study of the relations of living and extinct faunas as elucidating the past changes of the earth's surface* (Vol. I). London: Macmillan and Co.
26. Wandrei, K. (2017). *Examples of biogeographical proof for evolution.* Retrieved October 14, 2017, from https://sciencing.com/examples-biogeographical-proof-evolution-12944.html
27. Vuilleumier, & M. Monasterio (Eds.), *High altitude tropical biogeography* (pp. 317-373). New York: Oxford University Press.

2
CHAPTER

BIOGEOGRAPHY DISTRIBUTION PATTERNS

"Much of the geographical work of the past hundred years... has either explicitly or implicitly taken its inspiration from biology, and in particular Darwin. Many of the original Darwinians, such as Hooker, Wallace, Huxley, Bates, and Darwin himself, were actively concerned with geographical exploration, and it was largely facts of geographical distribution in a spatial setting which provided Darwin with the germ of his theory."

—David Ross Stoddart

CONTENTS

ABSTRACT

The study of different species and ecosystem in various kinds of geographic space, distinguished on the basis of geological time, is referred as biodiversity. All these species are distributed in a fixed biome and have a proper pattern defined. In this chapter, the biogeographical distribution pattern has been discussed in detail along with a brief description on the biogeographic realm of the ecosystem. There are many barriers in distributing the species in various patterns, which is also discussed below. Biographical distribution patterns also offer very interesting knowledge of the spread of biodiversity, hence approaches and methods of the distribution pattern of biogeography are also discussed in detail.

2.1 INTRODUCTION

Nature has allotted its space to organisms for living in a certain way. This organization is called the Distribution Pattern, and it is basically a territorial classification of where a particular species is to be found naturally. It is closely linked with the evolution of species over time and how such changes over time have emanated at the ancestral level. The modification of characteristics is a step-by-step process that is started with two main factors that are Reproduction and Environment. These two are proportional to the survival conditions that are also influenced by the inherited features. The pressure for adaptation can result from numerous factors. The adaptive features combined with habitat conditions then result in significant results affecting the species distribution.

The seasonal and permanent migration of organisms is sometimes responsible for varying distribution patterns, and their study can be complex and interesting at the same time. The climatic drivers can be high and low temperature extremes along with changing precipitation patterns. As we saw in the first chapter, the current distribution of biomes is controlled in principle by climatic factors; therefore, it is expected a variation in local ecosystems as long as environmental factors change over time given climate or land-use changes.

Another factor that is included in the distribution pattern study is the concept of "scale." Two similar species with little variations can occupy two different habitats in the same geographical region. For example, one bird species has its nest in thick branches while the other might have it on the forest floor. There can also be differences in micro habitats influencing spatial patterns of tree seedlings in thc forest floor or in savannas. On the other hand, it is important to find out if "barriers" are existent in this context as they play a constructive role. They can be climatic or topographical barriers, but also it is possible to find functional barriers. Increased involvement of humans can also function as new makers of spatial obstructions as can be for example the dams that create an obstruction for the free migration of fish upstream affecting their reproductive habits. There might also be overlapping factors which interactively affect the distribution patterns that can be analyzed with the help of multivariate statistics. There can be many limiting factors, and picking out the one responsible is not an easy task. It can also happen that one species becomes so prominent in one area that the other one might have to leave. This also affects distribution patterns, and it is an ecological principle of the ecological theory of communities called *competitive exclusion.*

The distribution of living beings through the biosphere is not uniform neither in time nor in space and each species, genus, family, order (each taxon) occupies a certain surface spontaneously, continuous or discontinuous that constitutes its territory, its "area of distribution." The distribution areas of the taxa are not random and evolve over time as the various environmental factors (i.e., relief, soil, climate, interspecific relationships, human presence) do, or depending on the biological evolution of the species. Its study, of the highest interest, is part of the framework of the chorology science that refers to the spatial distributions on earth. Each taxon has its particular geographical range given that it is determined by its own characteristics, adaptability, diffusion, and history.

2.2 MAJOR DISTRIBUTION PATTERNS

The distribution of plants and animals has certain patterns, which can be recognized to have a historical and geographical connection. This kind of inference approach leads to varied patterns of the organisms found in the whole world. Many tools like statistics are deployed by the research scientists use to gather proof of such connections. Results show a series of geologic and climatic factors identified to certain species in certain areas.

It should also be noted that different plants and animals evolve differently because the generations before them were settled in different places around the world. The same can be inferred from the history of islands. Some of them are the only habitats of a few bird species, fruit flies and unique category of snails. The biogeographic research of this fact from an evolutionary perspective can actually be a very complex study. Different scientists hence adopt different methodologies for the same. An example of this can be the existence of "Finches" in Galapagos Islands which are also found on some other islands.

Figure 2.1: A House Finch, a bird with a short forked brown tail and brown wings.

Source: https://www.allaboutbirds.org/guide/House_Finch/id

It is sometimes argued that the reason for this finch pattern of distribution may be the ability of birds to fly to a neighboring region, but then the same does not explain the existence over longer very long distances. The skeletal structure of some animals from different regions is considered to be a result of random mutations and existence of same environmental conditions elsewhere. Similarly, it is really interesting to observe that different regions have different stories of their flora and the fauna. Next, the three major categories of the distribution patterns before outlined will be detailed one by one.

Among the most common distribution types of areas are:

2.2.1 Zonal Distribution Areas

They are distribution areas of those taxa that extend forming a ring around the planet coinciding with a precise climatic fringe. They can be Circumboreal when they coincide with the high latitudes of the Northern Hemisphere as in the case of the plant family of the Betulaceae. Circumtemperate, Circumtropical, like the families of the Areacaceae (palmas).

2.2.2 Regional Distribution Areas

These are smaller surface distributions areas of those taxa that occupy a determined region of the world (usually between 10,000 and a few million km^2). Accordingly, it is said that a species can have a "Mediterranean," "European," "Andean," "Amazonian," "Patagonian ," "Guiana Shield," "Pantepui," and so forth.

2.2.3 Cosmopolitan distribution pattern

It is called as cosmopolitan the taxa that occupy the habitats that are favorable to them in all (or most) of the continents and oceans of the world. The phenomenon is very rare at the level of the species although more usual at higher taxonomic levels such as genera, family or order (in fact, meaning, strictly speaking, there are no absolutely cosmopolitan species with the current exception of the human being, so that cosmopolitan consideration extends to those that appear very extended by several continents or oceans).

Of the 160,000 known higher plant species only 25 species colonize more than half of the earth's surface. Cosmopolitanism appears mainly in aquatic or coastal taxa that benefit from a very homogenous medium (water duck is an example). On the other hand, there is a cultural cosmopolitanism that includes all the organisms that have been dispersed artificially by their preferential value as ornamental plants or pet animals (dog, cat, numerous garden plants …), for their agricultural interest (livestock, cultivated plants …) or that have accompanied spontaneously to the human groups for benefit of their activity (i.e., sparrow, rat, lice, nettle, diverse "weeds,").

"Cosmopolitan" sequence is the one in which the organism is found almost everywhere globally. We, as "Homo sapiens," are the major evidence of this category. Another example is *Falco peregrinus* and plants from genera like the Asteraceae *Senecio* and the fern *Pteridium*. This distribution is actually very wide, but certain species may have habitat or substrate preferences. The lichen *Psora decipiens* is one such species that thrives wherever there is calcareous soil. It is definitely found in many regions of the world, where such type of soil exists. Some types of species also have less specific habitat preferences. A species from different assassin bugs insect genera of "Peiratinae" is called to be the perfect illustration of the cosmopolitan genera. This bug is mostly found in tropical areas. Their daily diet is the other insects on which they attach at night.

Figure 2.2: An Assassin bug.

Source: https://www.flickr.com/photos/artour_a/2440280297/lightbox/

2.2.4 Disjunct distribution pattern

The Disjunct, as the word indicates refers to confinement in two or more separated areas. It can have possible causes like seed dispersal. Two major reasons responsible for this pattern are Vicariance and dispersal.

Vicariance is actually the breaking of a species' range. This results because of areas having a continuous distribution like land, tectonic plates, and mountains. For reference, consider the splitting of a region called Gondwanaland into continents. Also, this can result from the extinction arising through the interference among several populations. A pattern is termed as vicariant only if it has led to a continuous distribution.

These areas can be continuous or discontinuous ("disjuncts"), traits that provide us with information of interest on the history or dynamism of the taxon of which to be concerned. The continuous areas are those taxa whose populations are close enough to each other that could both interact with each other. They are most common in species of recent or expanding origin and should in principle be the most "Normal." Disjunct or discontinuous areas are those that appear fragmented in several parts (at least, since the question of limits could result problematic, so separated as to prevent contact between different populations).

The existence of disjunct areas is not always easy to explain. This is usually due to the fractionation of the initial area or the migration of groups of individuals. The distribution of a new species is continuous (it appears at one point and extends from it but "without losing territory" behind it). However, tailor-made over time, various natural or human-origin circumstances may lead to fragmentation. Once the area has been divided, the populations of each zone continue to evolve independently and may end up giving rise to new species ("allopatric" speciation process).

The Dispersal of species can occur if there are no suitable conditions for survival or reproduction. This can also cause dispersion over long distances. A good example is a deadly volcanic eruption of Krakatoa Volcano in 1883 which wiped out the living beings in that area. But in the next 50 years, the region saw life again, and it was the vegetation and animals coming from Java and Sumatra. The dispersion can also result due to mediums like forest corridors, river corridors, ocean currents working as corridors, filter bridges and so on. The means of dispersion shows a great variation depending upon a particular range or group of species.

It is also possible that the dispersal and the vicariance occur due to the same ecological event. One such example is the making of xeric savannahs that gave rise to forest expansion, after the disturbance of the forest ecosystems in East of Africa. A peculiar type of disjunct distribution that catches the eye of maximum researchers is the type in which the same type of organisms live in northern and the southern hemisphere. Initially, this was referred to as "Bipolar" distribution. A good example is the *Rhododendron* plant. The understanding of these distributions requires an organized system.

Figure 2.3: The Rhododendron plant, found in Asia, America and Europe.

(Source: http://rhodygarden.org/cms/our-plants/rhododendrons/about-rhododendrons/)

2.2.5 Endemic Distribution Patterns

Endemic refers to those taxa that appear strictly located in a reduced territory and are not found naturally in any other part of the world. In practice, however, the expression applies to taxa occupying widely varying surfaces, from a single site to a country. The concept of endemism is often used incorrectly, although there are no generally accepted limits when referring to an Endemic species should not be passed on the scale of a massif, region or archipelago, while when talking about genera or families, much categories are even more extensive, it can even reach the continent (as when it is said that cacti are endemic to the American continent).

Endemic can also be defined as the distribution specific to a particular area. An example could be a plant that grows in certain kind of soil only. An extension of this is *paleoendemic*, whose current population is the one left of a previous range of species. Some studies have led to conclusions that the regions with good plant endemism are also the areas with good animal endemism. That is to say, if a particular kind of plants abounds in one area and only that area, it is likely that this region also has animals that are not found elsewhere. This can be more common for small areas. This conclusion has been drawn from the nature of evolution as it occurs in all parts of the world. According to this assumption, it has also been noted that the dispersal can occur anywhere, and it is not restricted to some areas only.

It has also been found that as we increase the geographical area under consideration for Endemic distribution, a lot of complications arise. It becomes increasingly difficult to specify bigger areas as endemic.

The Mediterranean Sea and its surrounding lands are home to innumerable endemic groups. This region has a very high concentration of species specific to this area on account of which, it is said to have ten distinct hotspots of plant diversity. Several reasons contribute to the vast diversification in this area, and they include climatic, historical, and paleogeographical factors.

Figure 2.4: The Maritime Alps, Italy.

Source: http://www.recharge-green.eu/participatory-energy-planning-in-the-maritime-alps/

The Maritime Alps comprise 3,615 taxa, of which 115 are only found here and nowhere else. This widespread vegetation also leads to heavy competition that affects the ecosystem and the variety found in it at varying altitudes. About 60% of the endemics are found in rocky areas as compared to around 40% in higher altitudes.

The fauna of the islands in the ocean usually comes from the closest continents. As a result, the geography plays a big role in the distribution of the animals. The ecological conditions on the islands play an interesting role to harbor certain species. An ideal example is the stability and the insulated nature of the atmosphere in these places, which make them the home of certain extinct species as well. The animal communities that are more capable of dispersal and creating settlements are the first ones to migrate. This creates a competition amongst some and can cause a more divergent fauna that evolves greatly over time. These kinds of discoveries were actually responsible for inspiring one of the greatest Naturalists—Charles Darwin. His visit to Galapagos Islands led to the identification of many patterns of evolution. Some of them were alterations in body dimensions, decrease in dispersion inducing factors, modification of the populated areas and the reproduction changes.

The Galapagos Islands are widely famous over the globe. Spread over an area of about seven-hundred-and-seventy-kilometer square, they are 13 in number. The area is devoid of regional amphibians largely due to the fact that the climate is of arid type implying very low precipitation. Below is the list of the most common fauna occurring in this area:

- Land snails
- Fur seal
- Semi aquatic rodents
- Vespertillionid bats

Another example is found inland at South America Guiana Shield region known as the Tepui or table sandstones mountains that rise up to 2,700 m a.s.l. as 50 summit islands in a matrix of humid forested landscapes with present rainfall values up to 4,000 mm. The evolution of flora and fauna in Pantepui has attracted the attention of many researchers who found more than a third percent of the around 15.000 plant species are endemic. But also 35% of the endemic plants are only known from one of the tepuis. Plant species from the genus Heliamphora, Brochinia, Bonnetia, and many Orchids are among the peculiar endemic species.

Figure 2.5: The Lost world of the Tepuyes in South America, Venezuela.

Photograph from Jennifer Raga

Biogeographical classification of the world

The biogeographic classification of the planet is a difficult task since, as due to the high importance of endemics, a large part of the vegetation is ignored, and a greater knowledge of the evolution of plants is necessary to be able to correctly delimit all territories.

The ranges or hierarchies commonly accepted by biogeographers are: Floral Kingdoms or Faunal Realms The Floristic kingdom, is like an empire (floral kingdom, kingdom vegetation or floristic empire): it is the rank of greater importance, usually has a continental extension and for its division is considered the geological history of the land and the evolution of the fauna and the flora, mainly. The main centers of origin and diversification of flora and fauna at a global level are considered. It is after followed by Region, Province, Sector, District and finally the Tile as a basic unit.

2.3 BIOGEOGRAPHIC REALMS

A biographical realm can be defined as a portion of the earth, which possesses almost same ecological conditions. It is also called as an ecozone. The earth is divided into eight such realms as described previously. These are considered to be the major parts on the surface of the earth where diverse plant and animal life appeared and evolved. Each one will be considered in detail now. The biogeographical regionalization includes seven biogeographical "Realms" when divided by region as was defined by Wallace.

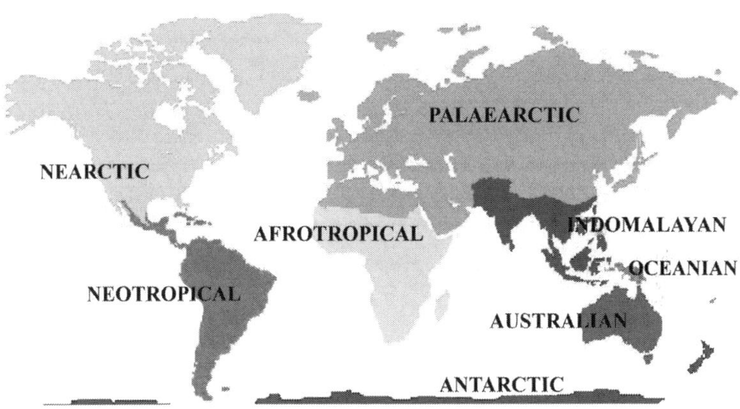

Figure 2.6: The eight Biogeographical Realms.

Source: https://www.google.co.in/search? biw=1366&bih=637&tbm=isch &sa=1&q=8+biogeographical+realms&oq=8+biogeographical+realms& gs_l=psy-ab.3…115476.115672.0.115960.2.2.0.0.0.0.0.0.0.0....0…1.1.64.psy-ab.2.0.0....0.bCGQIFpmjo4#imgrc=-sbJLgzrmS2nLM

The classifications can change if they consider phytogeographical realms or a combination of zoogeographical and phytogeographical regions.

Based on Wallace classification the Table 2.1 there are seven more recognized realms:

2.3.1 Palaearctic Realm

In terms of the size, this one is the largest of all. It covers the following areas: Europe, Central parts of Arabian Peninsula, and Asia, that is, Himalayan foothills and also Northern Africa. The natural environment of this realm is composed of multiple types. These include tundra region, taiga whose coniferous tress remain covered in snow, polar areas having algae, half grassland wetlands and mashes, coniferous forests, and broadleaf forests. Prairies, rainforests, deserts and coastal reefs are also a part. This realm has four distinct sub-regions as follows:

1) European Sub-region: Northern and central Europe, Black sea, and caucasus are included in it. It is represented by 85 families of vertebrates. Amphibians and Reptiles are represented with six families each. Myogale, only one genus of mammal is present. Bird like Tits, wagtails, mammals like wolf and moles are common in this sub-region.
2) Mediterranean Sub region: Remaining parts of Europe; Africa and Arabian portions are included in it. 124 families of terrestrial vertebrates are present. Birds like Upupa and Pastor, mammals like elephant scurew, Hyaena and porcupine are seen in this sub-region.
3) Siberian sub region: Northern Asia, Himalayas are included in it. 94 families of vertebrates are included in it. Families of Musk deer and Moles are confined to this sub-region.
4) Manchurian Sub region: Mangolia, Japan, Korea, and Manchuria are included in it. Mammals like, Tibetan Langur, Great Panda, Tufted deer, Chinese water deer are common.

One of the bird species found only here is the ancestors.

Figure 2.7: An Accentor bird from Siberia.

Source: https://www.google.co.in/search? q=accentors&source=lnms&tbm=isch&sa=X&ved=0ahUKEwilwPnMm-jWAhUGPY8KHaT1AtAQ_AUICygC&biw=657&bih=618#imgrc=fpeglw-mrJV8MM

A few other bird families found in this region are auks and waxwings. A few mammals include hamsters, rabbits, hares and red pandas. Other species found here are brown bear, reindeer, elk, and red deer.

2.3.2 Nearctic Realm

The part of North America up till the middle of Mexico is covered under this realm. The weather is of arctic type, and the center of the area is grassland, the western region has mountains and forests. The main fauna of this Realm is shown in Table 2.1 below:

Table 2.1: The Fauna of Nearctic Realms

(Source: Brittanica)

Mammals	Pocket mice, cats, bovids
Birds	Turkeys, humming birds
Reptiles	Alligator, horned toad
Amphibia	Tiger salamander, alpine newt, bull frog
Fish	Carps, perches, paddlefish

This area is further divided into the following sub regions including Californian, Rocky Mountains, Alleghany and Canadian. The Californian is a narrow region stretched between Sierra Nevada and Cascade Range, and from Vancouver to British Columbia. Free tailed bats are common here. The Rocky Mountains lie in the eastern region of California, and they have animals such as mountain goat and prairie dogs. Next is the Alleghany region that is the eastern part of the United States. Last is the Canadian that includes Alaska, Canada, and Greenland.

This realm has a vivid topography starting from continents, Islands extending to desert areas and polar areas. Figuratively, it is said to be running along the Arctic Circle. There are mountains with glaciers and icy islands in the middle of the frozen sea. The extreme south of this realm is the main continent, and this is shortened to an isthmus that is actually a barrier between this and the neotropical realm. To the north, there are seaways. In this region, the split and union of the touching parts of the continents have influenced the migration of animals. There are hilly regions and plain regions. Cold habitats exist here which are habitable for only a few months of the year. There is a strip of coniferous forest close to the area that is connected to the Palaearctic realm, and it has the same animals than those found in the palearctic realm. Many of the grazing animals are found here, and some of them have evolved from "sprintosaurs," descendants of a family of dinosaurs. They are very interesting. Their flag like tail helps them to confuse the predators when they move in groups in the grasslands.

Figure 2.8: Sprintosaurs.

Source: http://speculativeevolution.wikia.com/wiki/Sprintosaur

2.3.3 Neotropical Realm

It starts from the Mexican desert in the South of America and extends through Central America. Amongst plants, there are known species as the ornamental ones, rubber trees, and agaves. The animals seen in this region are jaguars, peccaries, tapir, insects and a large variety of birds. The most prominent bioma of this realm is the Amazon rainforest, which is a belt from the Andes Mountains to the Atlantic Ocean. Certain migratory birds that are found within the birds' communities have their breeding season in summer in Canada and the United States. They travel in winter to Mexico and Central or South America. It is worth knowing that the Neotropical Realm is the habitat of about 1,145 threatened amphibians. As expected, tropical rainforests are a characteristic bioma for this realm.

The tropical land covering the north of South America is mostly grassland. There are deserts as well in this realm. This is due to zero rainfall here resulting from their location in the rain shadow zone and the cold ocean currents. To sum it up, main ecosystems of this realm are besides tropical rainforests, grasslands, savannas with trees types of cerrado and cerradao in Brazil, dry forests, deciduous and semi-deciduous forests, mixed woodlands, desert scrubs, sclerophyllous shrublands, cloud forests, igapo and varzea flooded forests, *Mauritia* palm gallery forests. The Neotropical Realm actually has the need of concrete conservation measures. The knowledge of the functions of an ecosystem integrated with its biodiversity is extremely essential for uncovering the conservation strategies as we will see in the biodiversity chapters.

2.3.4 Afrotropical Realm

The Afrotropical Realm was previously famous as the "Ethiopian" zone. It includes areas from the Middle East to South Africa. The following regions can be found: Islands in Western Indian Ocean, South and East of Arabian Peninsula, South of Sahara Desert, Iran and southwestern Pakistan. It is safe to say that this realm means almost the entire continent of Africa. According to the measurements, the area figurately amounts to 22.1 million square kilometers. There is an intertropical zone close to the equator, and this is a stretch of tropical moist broadleaf forest. The Arabian Desert is also included in this realm.

As can also be observed there are many oceanic islands making up this realm. The Seychelles small Archipielag of islands in the Indian Ocean has a wide variety of endemic species like Gardiner's Tree Frog (*Sooglossus gardineri*), Aldabra Giant Tortoise (*Aldabrachelys gigantean*), and the Seychelles Scops Owl (*Otus insularis*) that for many years was considered extinct until some populations were found. This area also has the Mascarene Grass Frog, which is not an ancient endemic like most species, but it was brought to this area by human intervention. The breeding habits of the organisms are the usual ones. However, some habitats, like those of Seychelles amphibians, are declining.

Figure 2.9: Mahe, Seychelles: An island in the Indian Ocean.

Source: http://en.seyvillas.com/html/mahe-beaches

2.3.5 Indomalayan Realm

Also known by the name *Oriental*, this realm runs over 7.5 million square kilometers. Countries covered by it are India, Southern China, Indonesia, Philippines, Taiwan and Japan's Ryukyu Islands. The whole is further split into three areas spread naturally called the bioregions, and they are Indochina, Indian subcontinent, and the Sunda Shelf bioregion. Since a large part of this realm is Asian, plantations are the most common.

The fauna flourishing in this area is quite wide. It includes Asian brown tortoise, hornbill, bluebirds, and barbs. Among others are pigeons, pythons, Sumatran Tiger and tentacled snakes.

Figure 2.10: The Sumatran Tiger.

Source: https://www.worldwildlife.org/species/sumatran-tiger

Sharks and gibbons can also be seen here. Commonly called as the Indo-Malaysian region, it is home to arboreal and scansorial animals. They include gibbons, elephants, rhinoceroses, hedgehogs, antelopes, and blackbucks. There are rodents, flying rodents, dolphins, and domesticated animals as well.

Figure 2.11: Apes, called as Gibbons.

Source: http://news.nationalgeographic.com/news/2012/08/120823-gibbon-song-opera-singer-helium-science-environment/

2.3.6 Oceanian Realm

It is mostly made up of thousands of islands and is majorly the Central and the South Pacific Ocean. A large part of this ream is actually under water. As the name suggests, the maximum surface area is oceans. The Australian continent rules in this realm. Some nations that are also a part of this realm are New Guinea, New Zealand, Melanesia, Micronesia, Polynesia, and Hawaii.

Broadly we can divide this realm into three sets—the continental, high and low islands. The composition and the strata of the texture are different in all islands. The physical topography of these land surfaces is also an interesting study. Researchers have identified several ways through which the organisms reached these islands during glaciation like the travel across the ocean from South of Asia. An important characteristic of the species found on these islands is that some of them have been found to evolve from common ancestors. Also, this realm has a high collection of endemic genera.

The travel of plants through waters is assumed to have occurred as a direct consequence of the ocean currents in which birds also played a part in the dispersion. Some plants have even developed an interesting mechanism to stay in the air for longer times, and this includes the formation of spores. Trees like banyan, eucalyptus, coconut palms, and mangroves are often seen. Birds form an important part of the biodiversity in this region, and there are kiwis, seabirds, emus, and wekas. There are more than a hundred species of bats itself here, and some lizards include goanna and skink. Then there are mammals that lay eggs. The duckbill platypus and echidna are two of them. Koala, Kangaroo, and wallaby are also seen.

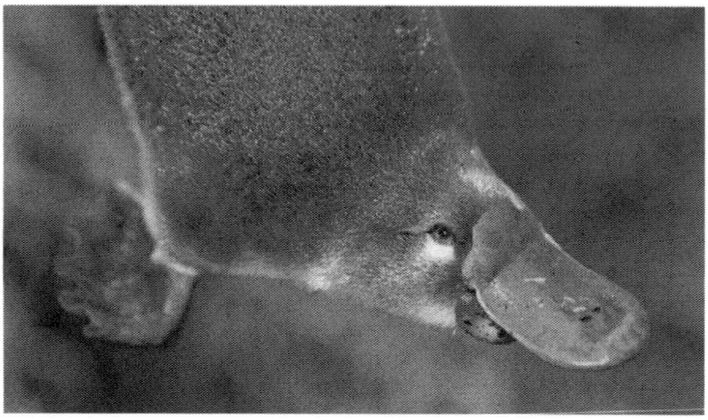

Figure 2.12: The Duck billed platypus, found in eastern Australia.

Source: http://news.nationalgeographic.com/news/2012/08/120823-gibbon-song-opera-singer-helium-science-environment/

Figure 2.13: Achidna, the Spiny Ant-eater.

Source: http://picssr.com/tags/achidna

The Marine Flora and Fauna is very rich in the Oceanian realm. Collectively, it contains the following aquatic realms: temperate Australasia, Central Indo-Pacific, and Eastern Indo-Pacific. The Flora and fauna share a lot of common features here due to the nature of the water environment. Seabirds of this realm found in Australasia attract many tourists from around the world every year. Coral Reefs are the most important ecosystems and are spots for tourism for a total off the board treat to the eyes from beneath the seawaters. Both of these are formally considered as underwater biodiversity hotspots. A wide range of sea animals is to be seen here. They are molluscs, crustaceans, sea turtles, whales, dolphins, and porpoises.

The climate of this realm is largely subtropical, having humid and dry seasons. These give rise to broadleaf forests, subtropical grasslands, savannas, and shrublands. The water as a dispersion medium is not very effective. There are no predators and grazing animals. Birds and bats are seen though. The onset of European people is said to have introduced species like cats, sheep, goats, and mongoose to this realm. Practices like hunting and forestation outside permissible limits have led to the endangering of certain species plus also the extinction of some. The human interruption has also affected the ecosystems here, and in some parts, winds have caused the conversion of a forest into a grassland.

2.3.7 Australian Realm

This realm is divided into sub regions or tropical biomes of savannas and deserts. The tropical biomes have dense forests with trees of ash, gum, and eucalyptus. The fauna comprises koalas, platypus, flying foxes and lyre birds. On a broader scale, a range of herbivores, carnivores, and insectivores abound here. Recent immigrants include rodents, and there is the spiny anteater among others.

2.3.8 Antarctic Realm

To the south of the Antarctic Ocean, this realm is located with cold climate almost throughout the year. The location of this realm is actually the meeting point of the subtropical waters with the cold polar waters. Antarctica, sub Antarctic islands, and southwestern New Zealand collectively make this realm. Freshwater snails are a common sight here.

Figure 2.14: Antarctic krill, known to be the only animal of krill size to directly use the minute phytoplankton cells.

Source: https://en.wikipedia.org/wiki/Antarctic_krill

Because of the cold climate, only a few plants have been able to thrive in such difficult conditions. They are numerous species of lichens, mosses, algae, and liverworts. In the peninsular area, Antarctic hair grass and pearlworts live. Animals like penguins, seals, and whales are the major ones found here. Some of the nearby islands in the Antarctic Ocean have tundra vegetation as well; the reason is the mild climate owing to their geographical location. One species that has caught the attention of many scientists in this area is the Antarctic krill. Many animals like squids, crabeater seals, leopard seals and so on prey this upon. Some researchers have also discovered the existence of microorganisms in as deep as 2,600 feet in the ocean.

2.4 BIOGEOGRAPHICAL BARRIERS

To make it simple, let's say that a barrier as generally defined as an obstruction to some process or objects. In terms of biogeography, they are the geographical barriers on the surface of the earth and their inner components that collectively affect an ecosystem and the variation of its biodiversity. Some factors affect only the plant species and other only the animal species, but there are most of them affect both. The primary barriers are large physical features like rough terrains, mountains, large rivers, snow covered regions, dry areas among other. Climatic conditions, which are interlinked with the species making a particular region constitute barriers.

Barriers to biodiversity have also come from human activities as can be extensive agricultural landscapes.

To get a clearer picture of what the barriers imply in a biogeographical context, let us consider an example. Waterfalls in mountainous regions alter the passages of movements for the aquatic organisms. The creation of water conveyance systems in dams can also be considered in the same context.

Figure 2.15: Desert Barrier.

Source: https://www.slideshare.net/lschmidt1170/ch-5-biogeography

It is also worth mentioning that the nature and the scale of the barriers also matter. The natural barriers like mountainous regions and oceans result in the formation of specific habitats that have conditions suitable to the animals and plants of that habitat. The isolation of the habitats can be a result of the combination of natural and manmade factors. In some cases, the construction of dams has affected the seasonal migration of certain species. The freshwaters have three major kinds of barriers that can be defined as continental, interbasin and within the basin. The separation of some fauna can also be due to the inability to cross-difficult formations of earth's surface like the high mountain ranges and the oceans. In many areas, it has been observed that some species of freshwater organisms didn't occur before human disruptions and explorations. An example is the mosquitofish that is now found in almost all zoogeographic regions.

In the case of basins, an ideal scenario would be the movement of fishes along the coastlines in an attempt to grow settlements in adjoining areas.

Glaciations have also resulted in the swiping of certain species in some areas and their further colonization in further areas in accordance with the water flow occurring according to the terrains. Shipping canals are yet another instance of the artificially made barriers affecting the biodiversity. It must also be noted that certain species can also be intentionally transported like for example the ornamental fish, the orchids, parrots.

2.5 APPROACHES AND METHODS OF DISTRIBUTION PATTERNS

Apart from the scientific point of view, biogeographical distribution patterns are significant to understand the evolution of biodiversity and ecosystems. Consider the starchy crop potato that is said to have originated from Peru and is now a staple food worldwide. Humans are the medium in this example, and there can be many other mediums for dispersal. The dispersal process of a species takes place within limits. This is the reason why we have different patterns of biogeographical distribution.

The never-ending alterations of the environment account for extinction on several grounds. They include the expansion of invasive species groups, decreased immunity because of the changing climate and pollution. A barrier like a river might not be an obstruction for all the fauna. Fish species and aquatic plants can still be transported to other regions. The abiotic factors associated with an environment are responsible for setting the range of some species. Each species has different characteristics that will allow them to expand their range. The identification of such traits is a topic in different animals and plants, which has a very wide scope for research. Furthermore, analyzing the ways through which the specific traits can also be cumbersome as the success or failure cannot be predicted beforehand without taking a lot of observations. The range limits have been found to be largely affected by the climatic factors that change according to the elevation levels. Treefrogs of Mexico live in the northeastern region, but they cannot move to temperate regions because of their intolerance to cooler winters.

The study of the relationship between a prey and its attacker also affects the distribution patterns of some of them. The body size of an animal can also affect the distribution of the predators. When the differences in size are big, it causes different predators to have different range limits. Sometimes, the inter-species interaction gives rise to large-scale geographical patterns. The intervening regions between two realms do not have very definitive boundaries but have favorable climatic conditions for the reproduction amongst two kinds of species. It is also sometimes argued that the species interaction doesn't always lead to the formation of certain patterns. The example of islands can be cited in this regard. There is very little evidence to prove whether any region has become saturated with its native plants. At times, even the natural predators can cause the swiping of a species from a particular area. The detailed case studies of certain behaviors of species also put forth many interesting implications and also a lot of room for future research. The similarities of species in communities can also be used to extract inferences of ecological processes.

The study of biogeography actually brings us closer to the patterns of nature as well. Varying methods adopted by different researchers have also led to disagreements among naturalists, and hence the patterns have become an even hotter topic. The debate over the fundamentals has been quite a thing. The time-calibrated calculations are very effective in examining the history of the patterns.

The overlapping and replacement of geographical regions affect the spatial patterns of the biodiversity. The related patterns of variations give rise to many questions as well. The colonization of the isolated regions is also another interesting field. Many biogeographical distributions have had to be revised at times due to the continually occurring change in the geographies. Poor quality of water bodies is sometimes responsible for obstructing the expansion of non-native species. One such case is that of the Chicago Sanitary and Shipping canal that was once a dumpsite for the city's sewage. As its importance as an ecosystem that joins the Great Lake and Mississippi River basin was recognized by biogeographers, efforts were made for cleaning treatments and eventually the waters became suitable for fish faunas of the region. Similar patterns occurred in the Thames in London UK.

The restoration of natural habitat conditions can be done as a measure toward the preservation of species, which could be endangered due to multiple reasons being habitat loss the main of them. It has also been found that the selective breeding of domesticated animals can bring about changes in the same way as the natural changes over a period of time. Earlier biogeography was not a separate topic with its own identity, but the increasing human negative impacts and influences over the biological resources of the earth have led to its existence as a complete subject in itself now being an important part of the environmental sciences. The alarming rate of the consequences of human actions and the awareness about its impact on the natural environments are the major cause. The credit of many discoveries goes to the naturalist Charles Darwin, famous for his theory of evolution which he formed while he traversed on a five-year long voyage.

2.6 EVOLUTION THEORY AND EVIDENCE

Darwin's theory of evolution is considered, with justice, as the greatest unifying principle of biology. Darwin was not the first to propose a theory of evolution, but he was the first to describe a valid mechanism by which it could occur. His theory differed from previous theories in that he imagined evolution as a double process, which depended:

1) on the existence of heritable variations among organisms, and
2) on the natural selection process by which some organisms, by virtue of its inheritable variations, left more progeny than others.

There is evidence that shows the existence of the evolutionary process. Distinguishing the field from which they come, five sources of evidence can be recognized: direct observation, biogeography, the fossil record, study of homologies and the imperfection of adaptation.

Since the time of Darwin, a large amount of new evidence has accumulated in all these categories, particularly at the cellular, subcellular and molecular levels, which highlight the historical unity of all living organisms. A central weakness of Darwin's theory, however, which remained unresolved for many years, was the absence of a valid mechanism to explain inheritance.

In the 1930s, the work of many scientists was embodied in the Synthetic Theory of Evolution, which combines the principles of Mendelian genetics with Darwinian theory. Synthetic Theory has provided—and continues to provide—the foundation of the work of biologists in their attempts to unravel the details of the history of life.

Charles Darwin was not the first to propose that the diversity of organisms is the result of historical processes, but recognition for the theory of evolution belongs to him for two reasons. In the first place, his "long argument"—as the Origin of Species was characterized—left little doubt that evolution had actually occurred and, thus, marked a turning point in the science of biology. The second reason, which is intimately linked with the first, is that Darwin perceived the general mechanism by virtue of which evolution occurs. The original concept of Darwin and Wallace about how evolution occurs still continues to provide the basic framework for our understanding of the process. That concept is based on five premises:

a. Organisms generatc similar organisms; In other words, there is stability in the reproduction process.

b. In most species, the number of individuals that survive and reproduce in each generation is small compared to the total number produced initially.

c. In any given population there are random variations between individual organisms, some of which are hereditary, that is, they are not produced by the environment.

d. The interaction between these hereditary variations, arising at random, and the characteristics of the environment determine to a significant degree which individuals will survive and reproduce and which will not. Some variations allow individuals to produce more offspring than others. Darwin called these characteristics "favorable" variations and proposed that favorable inherited variations tend to become increasingly common from one generation to the next. This is the process that Darwin called natural selection.

e. Given enough time, natural selection leads to the accumulation of changes that cause differences between groups of organisms.

It is, therefore, noted that the research into the distribution patterns is indeed a vast topic and that only a few topics of it can be touched at once. As we go deeper into the functioning of the ecosystems and its connection with its habitants, more and more dependencies arise. These, in turn, make it difficult to arrive at specific conventions for defining the patterns. One research topic is often found to be the gate of n number of possibilities enlisting the correlations. For example, it is easy to establish the origins of small-scale patterns, but it becomes increasingly difficult to do the same for large-scale ones. It is also very difficult to say that similar patterns exist for a full range of habitats. The factors affecting the diversity not only attempts to explain the distribution, but it has a lot of other aspects that throw light on more important things.

However, systematic approaches have always brought good results in the past and so has the biogeographical distribution patterns. Their study has been the major source to drive the conservation policies. Today, there exist extensively developed measures that operate through processes based on research and technology. But In spite of the advancements, there is still a long way to go due to the changing nature of the earth.

REFERENCES

1. Berry, P., & Riina, R. (2005). Insights into the diversity of the Pantepui flora and the biogeographic complexity of the Guayana Shield. *Biologiske Skrifter*, *55,* 145-167.
2. Casazza, G., Zappa, E., Mariotti, M., & Médail, F. (2010). *Distribution pattern and richness of endemic plant species in Maritime and Ligurian Alps*. Retrieved October 13, 2017, from http://www.bmib.it/documents/72/19%20casazza.pdf
3. Chiarucci, A., Beierkuhnlein, C., Storch, D., Weigelt, P. and Dengler, J. (2014). *Global patterns of vascular plant species richness, endemic richness and endemicity: a new approach to identify hotspots and cold spots*. In L. Mucina, J. N. Price, & J. M. Kalwij (Eds.), Biodiversity and vegetation: Patterns, processes, conservation: 78. Perth, AU: Kwongan Foundation.
4. Chłond, D., & Bugaj-Nawrocka, A. (2015). Distribution pattern and climate preferences of the representatives of the cosmopolitan genus *Sirthenea Spinola*, 1840 (Heteroptera: Reduviidae: Peiratinae). *PLOS One*, *10*(10), p.e0140801.
5. Huber, O., & Garcıa, P. (2011). Peatlands of the Western Guayana Highlands, Venezuela. *Ecological Studies 217.* doi: 10.1007/978–3–642–20138–7_3
6. Huber, O. (1994). Recent advances in the phytogeography of the Guayana region, South America. *Mém. Soc. Biogéogr. 4*, 53-63.
7. Ritter, M. (2012). *Earth biomes*. Retrieved October 13, 2017, from http://www.earthonlinemedia.com/ebooks/tpe_3e/biomes/biogeographical_realms.html
8. Rosen, D. (1978). *Vicariant patterns and historical explanation in biogeography*. Retrieved October 13, 2017, from https://academic.oup.com/sysbio/article-abstract/27/2/159/1614255
9. Rull, V., Vegas-Vilarrúbia, T., Huber, O., & Señaris, C. (Eds.). Biodiversity of Pantepui, the Lost World of the Neotropics. Cambridge: Elsevier (in press).
10. Rull, V., & Vegas-Vilarrúbia, T. (2017). Potential responses of vascular plants from the pristine "Lost World" of the Neotropical Guayana Highlands to global warming: Review and new perspectives. *Frontiers in Plant Science*, *8*, 81. doi: 10.3389/fpls.2017.00081; ISSN: 1664-462X

3
CHAPTER

BIODIVERSITY

"Biodiversity can't be maintained by protecting a few species in a zoo, or by preserving greenbelts or national parks. To function properly, nature needs more room than that. It can maintain itself, however, without human expense, without zookeepers, park rangers, foresters or gene banks. All it needs is to be left alone."

—Donella Meadow

CONTENTS

ABSTRACT

There is growing evidence that Biodiversity and its conservation are essential for the survival of humanity. Genes, species, ecosystems, and landscapes of a particular region are collectively representing its biodiversity. The biodiversity found on the earth currently is the product of around 4 billion years of evolution of living organisms. In this chapter, a brief discussion has been done on biodiversity concepts, its types, and the importance of biodiversity. Various causes of biodiversity loss and its patterns are also described in detail along with the practices required for its conservation.

3.1 INTRODUCTION

The biodiversity is a result of the evolutionary process that manifests itself in the existence of different ways of living beings. Mutation and selection determine the characteristics and the amount of diversity that exists in a given place and time. There are differences at the genetic level, differences in morphological, physiological and ethological phenotypes, differences in the forms of development, in demography, and in the life stories. Biological diversity covers the entire scale of organization of living beings. However, when we refer to it in a conservationist context, we are talking about diversity of species, variation intraspecific and intra-population, and ultimately genetic variation.

Biodiversity, also known as biological diversity, includes a variety of kinds of living organisms from various sources like, terrestrial and aquatic and it also comprises ecological complexes at different scales like genetic diversity, diversity amongst the species, and diversities pertaining in the communities and ecosystems. The foundation of any ecosystem is the biodiversity present on earth, where all the living organisms are inter-related and inter-dependent on each other. According to Singh, "the current estimates of the total number of species on earth vary from 5 to more than 100 million, with a more conservative figure of 13.6 million of species. Of these, only 1.78 million of species have yet been described and awarded scientific names." Each species is adapted to live in a specific environment, from mountain peaks to the depth of seas, from polar ice caps to tropical rain forests and deserts."

Apart from the ecological balances and its importance, biodiversity also provides socioeconomic and monetary support and an asset to a country. Human beings have always been dependent on the natural resources available in the ecosystems helping them to sustain their livelihood through the use of various goods and tangible or intangible ecological services available to them.

Regarding the historical development of biodiversity, we could say that the first classification of living beings was conducted by Linneo who classified all living beings in Domains, Kingdoms, Phyllum, Class, Order, Family, Genus and finally Species, using morphology and natural history or phylogeny. The term biodiversity was first coined by Walter Rosen in the year 1986, while he was planning for National Forum on Biological Diversity, which was organized by the National Research Council. The first in utilizing the term in the scientific literature, however, was Edward Wilson. In general, biodiversity could also be described as "variation of life at all levels of biological organization." It is also defined as the diversity of alleles, genes, and organisms, which study the processes related to mutation and gene transfer, which drive the evolution of a living organism. According to the Convention on Biological Diversity, "the variability among living organisms from all sources including, inter alia, terrestrial, marine and other aquatic ecosystems and the ecological complexes of which they are a part; this includes diversity within species, between species, and of ecosystems."

3.2 TYPES OF BIODIVERSITY

Biodiversity includes genetic variation within species, the variety of species within an area, and the variety of habitat types within a landscape. Biological diversity is of fundamental importance to the functioning of all natural and human-engineered ecosystems, and by extension to the ecosystem services that nature provides free of charge to human society. Living organisms play central roles in the cycles of major elements (carbon, nitrogen, and so on) and water in the environment, and diversity specifically is important in that these cycles require numerous interacting species. Some species are key pieces of the puzzle that represent the functional organization and equilibrium of an ecosystem.

Taxonomy is a term of high importance in biodiversity as it refers to the science of discovering, naming and classifying all the animals, plants and microorganisms across the world. Taxonomists are the people who arrange the different species into classification according to their genes, morphology, and behavior and biochemical observation. Classifying various species is important for the fundamental understanding of the biodiversity and the conservation of biodiversity. Taxonomy is essential because it is a language through which communication is possible as it is an important infrastructure for different activities in the society. For instance, if humans, animals, and other things do not have names, how would they communicate with each other. Human beings classify various things in the same manner as the taxonomist differentiates among species that is by assigning names to similar objects so that they are recognizable. In past centuries, taxonomists have found several species and named approximately 1.78 million known and discovered species of plants, animals, and micro-organisms.

There are different types of animals, plants, fungi, bacteria and other micro-organisms and all these types are known as species. Each species is different biologically. A species has the potential to produce descendants that can interbreed themselves. For example, animals of two same species can interbreed and produce offspring, but two different species of animals cannot interbreed. The role of taxonomists is to provide the unique name to species so that each species can be easily recognized and more can be learned about them. However, every country has a different language. Therefore, species can have names in many languages, but if a hedgehog is considered, then it is important to know which species of hedgehog is being considered. For this reason, all the species are classified on the basis of scientific names, which are a unique and universal identifier. The scientific names are universally accepted names due to which no confusion arises.

If taxonomy is poor, one could not recognize the different effects of different species. Moreover, with poor taxonomy, no research could be conducted on the funding of biodiversity study from the council of research. In addition to this, it is considered that taxonomy is an important infrastructure for different fields of research such as biotechnology, genetics, ecology, and physiology. Along with this with the help of taxonomy pharmaceutical researches and research on veterinary medicines can also be conducted. It is with the help of taxonomy that humans are able to discover several numbers of species but still, we, have no idea about the total number of species present on the earth as many species are still undiscovered. According to the global database information, there are about 1,160,711 species that are currently discovered. Poor taxonomy or lack of taxonomy is an impediment for the society.

There are various types of diversity of living beings on the planet. All biological and chemical processes are life sustaining. This rich variety of genes, species, biological communities, and food web chains in life-sustaining ecosystems provide food, fibers, energy, raw materials and medicines and help the world economy to grow. Free waste recycling and cleaning from wetland ecosystems, pest control services and purification of various raw materials are only some of the major services provided by the components present in the ecosystems. On the other hand, nowadays every species contains genetic information representing evolutionary processes along thousands to millions of years of adaptation to the earth's changing environmental conditions and its existence even if it is in low quantities is the raw material for future adaptations that allow survival of life in the earth.

If there is a loss of biodiversity, it will mean a reduction in the availability of services that are generally provided through the resources available in the ecosystem. Also, loss of biodiversity leads to a decline in the number of species available and also the communities. It becomes difficult for the ecosystem to adapt the myriad of disasters and major changes taking place in the environment. Biodiversity is also referred as the insurance policy that green governments would have to fight against any disaster. According to the preferences and understanding of various biotechnologists, human cultural diversity is also an essential part of earth's biodiversity as their interaction with local environments have led them to develop empirical knowledge about biodiversity uses and values beyond to material and that can also be considered from its spiritual and religious relevance. The variety of human cultures also represents numerous social and technological solutions to changing environmental conditions. Below a deeper insight will be included with regards to the different types of biodiversity.

The following are recognized as different types of biodiversity levels:

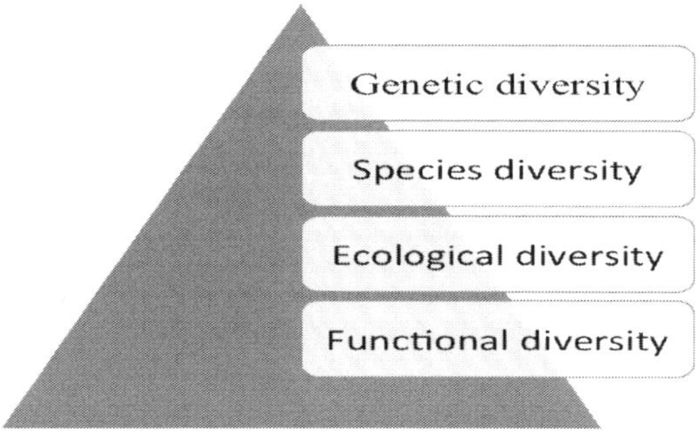

Figure 3.1: Types of biodiversity.

3.2.1 Genetic Biodiversity

Genetic diversity is the one where the species of plants and animals are differentiated on the basis of the genes they carry. Genetic diversity is defined as the total number of genetic characteristics in the genetic makeup of populations in a particular species. It is because of the genetic makeup of an individual, which makes it different from the other members of that particular species or community. This is because of a large number of combinations that are possible in a gene that provides every individual a different and a specific characteristic. It is distinguished from genetic variability, which describes the tendency of genetic characteristics to vary. Genetic diversity helps an individual in its adaptation to changes that occur in an environment. That as a way how populations are adapted to changing environments; according to genetic theory, when each of two or more alternative forms of a gene arise by mutation and are found at the same place on a chromosome, then there are chances of more variations which would be suitable for the environment. Such individuals are more likely to survive to produce offspring bearing that "allele."

The population will continue for more generations because of the success of these individuals. For example, it is very much visible and also understandable that every individual is different. For a healthy breeding population of any community or species, this genetic variability is very important. When a total number of breeding individual also diminish, there are chances of dissimilarities in genetic makeup that would result in its reduction. Hence, in-breeding occurs due to the reduction in breeding individuals. Ultimately, the species gets extinct. The Dodo flightless bird that got extinct and was endemic to the island of Mauritius, at the East of Madagascar in the Indian Ocean is a remarkable example.

The diversity in wild species forms the "gene pool" from which our crops and domestic animals have been developing over thousands of years. Therefore, it is possible nowadays to talk about Agrobiodiversity that according to FAO is defined as the result of natural selection processes and the careful selection and inventive developments of farmers, herders, and fishers over millennia. Today, the variety of nature's plenteousness is further harnessed by using wild relatives of crop plants to create new varieties of more productive crops and to breed better domestic animals. Genes are now getting manipulated through modern biotechnology so that medicines and other variety of industrial products could be developed in a better manner. There are different varieties of species of plants and animals available in natural ecosystems, for example, fruits, animals, fuel, wood, food and fodder, medicines, and so forth, whereas timber plantations do not provide a large variety of goods that are essential for local consumption.

The value of a natural forest ecosystem is richer than a timber plantation because the plantation has a lower diversity.

3.2.2 Species Biodiversity

Species biodiversity is represented in a defined spatial-temporal dimension resulting from a set of interactions among species that are integrated into a process of selection, mutual adaptation, and evolution.

Diversification of life follows tremendous differences at the cellular level, as Eukaryota (Animalia, Chromista, Fungi, Plantae, Protozoa) living beings differentiates from Prokaryotes (Bacteria and Archea). Cells in Eukaryota almost all cell division includes mitosis and eukaryotes present membranes surrounding their DNA, while Prokaryotes doesn't. Even though in Prokariotes the cell does not undergo mitosis, the end result is the same. The single chromosome is replicated, and the two copies split into two halves of a dividing cell. In these organisms, the process is called binary fission

The term diversity of species (or simply diversity) is often used to refer simultaneously to the number of species in a community and to the relative abundances of those species. That is, species diversity has two components: on the one hand, specific richness, which is the number of species in the community and is usually represented by the letter S, and, on the other hand, equitability: "Equitability" or "evenness"), which is a measure of the relative abundances of species usually represented by the letters E or J. The simplest index of species diversity is species richness (i.e., S). Species richness is simply the counting of all species within the community, but the quantification of species richness is not a simple task. In fact, it is unlikely that the total number of species inhabiting any community on Earth will be known although last estimates range from 1 million to 6 million of species on earth and higher including estimates of all variability found in bacteria species that can reach up to 70–90% of all variability followed by insects with plants representing less than 0.1% of all species variability on earth as can be seen in Table 3.1. Species richness, however, is not always linked to the importance of living groups as with the evolution of living beings in communities; complex individual organisms with asexual reproduction can be found dominating communities and ecosystems. Also, complex associations of species of viruses, bacteria, plants, and animals have coevolved to play a role in an ecosystem, and they are each counted as individual species, but they should be considered as functional living units.

Table 3.1: Estimates of species richness in different living beings? source?

SPECIES		**EARTH**			**OCEAN**	
	Catalog.	**Pred.**	**Sdev**	**Catalog.**	**Pred.**	**Sdev**
EU-KARY-OTES						
Animalia	953,434	7,720,000	958,000	171,082	2,150,000	145,000
Chromista	13,033	27,500	30,500	4,859	7,400	9,640
Fungi	43,271	611,000	297,000	1,097	5,320	11,100
Plantae	215,644	296,000	8,200	8,600	16,600	9,130
Protozoa	8,118	36,400	6,990	8,118	36,400	6,690
Total	1,233,500	8,740,000	1,300,000	193,756	2,210,000	182,000
PRO-KARY-OTES						
Archea	502	455	160	1	1	0
Bacteria	10,358	9,680	3,470	652	1,320	436
Total	10,860	10,100	3,630	653	1,320	436
TOTAL	1,244,360	8,750,000	1,300,000	194,409	2,210,000	182,000

There are a number of problems in accounting for diversity. One has to do with body size. It is relatively simple to locate and identify larger organisms: vascular plants, mammals, birds, reptiles, amphibians, fish and even the largest invertebrates. However, the diversity of microorganisms is a difficult task to accomplish. The second difficulty in quantifying the species diversity of a community is related to rarity. A few abundant species usually characterize a community, in which most species are rare. The combination of these two factors, size and rarity, with the net number of organisms that make up a community makes a listing of species an overwhelming task. On the other hand, it has been proposed that the simplest measure of the structure of the communities is the count of the number of species that exist within it, this mixture of species, in their number as in their relative abundance defines the biological structure of a community and is known by the name of species richness. However, within the set of species that make up the community, not all are equally abundant. We can discover this feature if we count all the individuals of each species in a series of samples within the community and determine what percentage of each contributes to the total number of individuals of all species. This measurement is known as relative abundance. The frequency is another variable to consider when conducting diversity measurements: it means the number of times that a species appears in a series of samples.

The analysis of the diversity needs to be considered in different spatial scales. Therefore, one has to consider different components. For example, it may be interesting to distinguish between two main types of diversity: α (also called local or taxonomic) diversity and γ (or regional) diversity. Local diversity is the number of species in a small area with more or less uniform habitat. It is clear that this measure of diversity depends on the definition of habitat that has been made, and the area and sampling effort. Regional diversity is the total number of species observed in all habitats in a region. When the same species are present in all habitats in the region, local diversity and regional diversity are the same. When each habitat has a unique flora and fauna, regional diversity is equal to the average local diversity multiplied by the number of habitats in the region. A third type of diversity is known as β (mosaic or motif diversity) and refers to the degree of change in species diversity along a transept or among habitats. Then the greater the difference in terms of specific composition among the different habitats of a region the greater will be the β diversity.

For the comparison of patterns of species richness and abundance among communities, a relative abundance chart of each species is generally used in relation to the range, where the range is defined by the order of species from the most to the least abundant. Then, the most abundant species will be represented first on the x-axis, with its corresponding value on the y-axis, which is the relative abundance value. This process is continued until all species are represented. The resulting graph is called the range-abundance diagram. Relative abundance has to do with the richness of species and the way individuals of each species are distributed in the range-abundance diagram.

The greater species richness is reflected by, a greater extent of the range-abundance curve (number of species), and the more equitable distribution of the individuals of each species (equitability of the species) is indicated by the more gradual gradient of the curve of range-abundance. Although the graphical procedure of range-abundance diagrams can be used to visually assess (interpret) differences in the biological structure of communities, these diagrams do not quantify the observed differences. Ecologists solve this need with the development of diversity indexes, which consider both the number and relative abundance of species within the community.

One of the simplest and most widely used indexes of diversity is the Simpson index, which measures the probability that two randomly selected individuals from one sample belong to the same species. Another index that is frequently used is the Diversity Index of Shannon–Wienner, in which it starts from the premise that the diversity in a natural system can be measured in a similar way to the information contained in a code or message. In fact, from the point of view of sampling, greater diversity represents a greater uncertainty in the identification of an individual drawn at random.

There is a diversity of theories to explain species diversity, and from them, we can understand that species diversity cannot be explained only for the physical environmental variation at a local scale, but that we can recognize several factors interconnected, including competition, dispersal, and predation. Variation in the physical environment and other factors also interact to determine these patterns of diversity through the combination of biotic and abiotic factors. Changes in species diversity (number, identity and relative abundance of species) in space and time is one of the oldest and best studied subjects in ecology. Despite this, there is little agreement on the underlying causes of many patterns of diversity. Understanding the causes of diversity implies: How does diversity arise? How is diversity distributed in space?, How is diversity maintained?. There are many theories that try to explain species diversity. Which one, if any, is correct and why is it important? The importance lies in the current movement to conserve biodiversity on the land knowing that what promotes diversity helps us to preserve it. As to which theory is correct, it is necessary to examine the most widely proposed theories and examine the evidence supporting them, that is why trying to explain the theories of species richness is to some extent illusory, since there are several theories that aim to increase our understanding or dominion over the richness of species diversity.

To determine the diversity of species, a series of patterns such as Taxonomic (that implies not only having reliable knowledge about the species that inhabit a given space, but also their distributions in that space), Spatial (latitudinal gradient, altitudinal) and Ecological (biotic and abiotic factors), we can divide the theories explaining species diversity along different regions into two main categories: biotic explanations and abiotic explanations. Among the explanations that argue for the increase of diversity that is more related to biogeographical explanations, we have:

A. Theory of spatial heterogeneity

According to Stiling 1995, spatial heterogeneity can be defined as the variation in space in the distribution of a point pattern, or variation of a qualitative or quantitative value of a surface pattern, which can be caused by habitat factors and their temporal variations. It includes the density of resources that can result in a series of appropriate patches of different sizes and levels of isolation, individual traits such as dispersal skills, sexual attraction by pheromones or aggregate behavior that have implications for population dynamics and distribution of species. Therefore, it is of great importance in the study of populations, communities, ecosystems, and landscapes how are the individual traits and the neutral processes operating in the different temporal and geographical scales.

We can exemplify a neutral theory with tropical forests, where it is common to think that they are monotonous, which is a big mistake, as they differ in terms of environmental and evolutionary history, as well as their regional and local climate, relief, soil, and hydrology. The largest scale on which species diversity is normally examined is the Region/continent (gamma diversity), through the so-called latitude-area effect, a clearer example is the rainforest, which is found in four major regions; Central and South America, Central and West Africa, and Indo-Malaya and Australia. In a single continent, more than 20 types of forests have been recognized in the Amazon, South America. Different Forest Types are important for raising diversity as many species are confined to some of the types between habitats or beta diversity, which is frequently studied along hydrological gradients.

B. Diversification of niches

Individuals of a given species can maintain viable populations only under a certain range of resource conditions, and for two similar species to reside in the same habitat, they must divide resources. If the resource needs of the two species overlap, resources tend to be limited locally, interspecific competition occurs, and one species will move the other. The two species will continue to compete within their tolerance limits. For example, one argument used for the tropics is that while species are more evolved, interspecific competition is more intense and, therefore, species occupy smaller niches than in temperate areas.

C Biotic interactions

These are interactions among species such as predation, herbivory, parasitism. Herbivory requires that predators, particularly seed-eating insects, be specific to the host and there is some evidence of this. When herbivory is locally prolific, it provides a key mechanism for maintaining a state of equilibrium and preventing rare species from becoming too common. A second important biotic explanation with biogeographic implications considers the theory of recruitment limitation. According to this theory, the local absence of superior young competitors in places favorable for their survival is due to poor dispersal ability, low abundance or fortuitous events. Although there is not a definitive competitive exclusion, the constraint of recruiting competitors works to maintain local populations indefinitely. A third biotic interaction deals with a temporary partitioning of resources: the temporal partition of resources refers to the differentiation between the moment of plant and animal activity. For example, tropical trees, whether dispersed by animals or by the wind, tend to show seasonal variation in the time of fruiting and seed germination. This causes a greater capacity of regeneration of the species for the following months and, therefore, among species, there is a seasonal variation in the advantage of the germination. Inter-specific competition is reduced, and this allows many species to coexist. Being food supplies plentiful, but competition is minimized. The temporal partition of the niche space also takes place locally on a daytime basis. For example, many species of small mammals (e.g., pigs) are nocturnal to avoid competition during daytime hours. The complexity of food webs in tropical forests makes it almost impossible to test hypotheses regarding the competitive advantage obtained from the temporal partition of resources. Temporary allocation of resources can alleviate competition for interference, but not to reduce resource competition.

D. A fourth is the theory of accumulated evolution based on observations made by the early naturalists mentioned before Charles Darwin and Alfred Russell Wallace. They assert that the rainforest environment would have low extinction rates, sparing the frequent interruption of winter and occasional disturbance of glaciations and thereby accumulate species. For them, diversity was explained as the product of short and long-term environmental stability that promoted the presence of biomes in the long term and allowed time for speciation and co-evolution (hence the alternative name of the stability hypothesis in the time). The cumulative evolution is reinforced by interactive co-evolution. Coevolution, the mutual adaptation of two or more species by complementary evolution, can influence the abundance and distribution of species. For example, fig species are pollinated by symbiotic species of wasps that can only be reproduced in female flowers of their specific host. This supports the continuity of species throughout generations. Figs are then involved in a two-phase dispersion process by vertebrates and ants, which ensure their distribution over space.

E. Theory of productivity, seen not only as energy production but also as the amount of energy captured and transformed into living matter per unit area and time. This theory proposes that greater production produces greater wealth; That is, when there is a broader base to the energy pyramid it allows more species in that pyramid. Thus, long-term productivity initially promotes the generation of viable species populations but also manifests itself in forest or lake habitats through the complexity of vertical stratification. It also incorporates improved niche processes.

F. Theory of the area, which is based on the idea that to a greater extent of an area, a continent, for instance, the possibilities of isolation among populations increase, with corresponding increases in the possibilities of speciation. It has also been shown frequently that larger areas support more species. Thus, greater areas of climate similarity will have greater species richness. Globally, there is climate symmetry between the polar regions and the temperate zones, but only in the tropics, we see adjacent symmetrical opposing climates, creating large areas. Most of the dominant species evolved in the larger areas and spread outward, creating a gradient of species richness. None of the theories, however, seems capable of explaining why, if wealth is relative to the area, there should not be more species in the vast mass of contiguous land of Asia. North America has an area larger than Central America but has far fewer species of birds and mammals. Neither the surface nor the shelving areas of the North Pacific are larger than those of the North Atlantic, but, however, the North Pacific has a relative richness of species in the trees, it is also limited in unproductive environments (such as north of Asia). Bleeding species from one area to another may obscure patterns. His idea is that these three elements, area, productivity and zonal bleeding cause most of the latitudinal gradients we see.

G. Genetic drift and isolation: According to Genetic drift and isolation theory of Hill and Hill, species diversity is partially dependent on generalized self-pollination and population inbreeding due to dispersed distributions of highly specialized plant species. Opposing vicarianism, it is increasingly recognized that speciation can also take place without geographic (sympatric) isolation. In the presence of microenvironments of heterogeneity and absence of extensive gene flow, sympatric fragmentation of populations is possible, which may lead to random genetic drift, Polymorphism and eventually to speciation. Speciation under these conditions, however, makes a much smaller contribution to diversity than large-scale allopatric speciation.

H. Rapoport Effect. Eduardo H. Rapoport had referred his study to a phenomenon, in which it suggests narrower environmental tolerances of tropical species in comparison with temperate species. If both groups were dispersed at the same amount, a larger number of transient temperate species would occur in tropical habitats and could live there for a relatively short time. It has been shown that the establishment of species at sites where they cannot sustain themselves occurs in some plant communities on a local scale. For example, marshes are found in temperate, but not tropical, coastal areas; Mangroves are found only in the tropics, not in the temperate zones. Thus, living temperate organisms of marshes could not survive in the tropics.

I The Intermediate Disturbance Hypothesis. According to Hill and Hill, this theory explains how species diversity changes as the disturbance levels change. An ecological disturbance can be anything that changes the characteristics of the ecosystem or a part of it, so it can be said that all ecosystems are continuously disturbed, although most of these disturbances are usually of small intensity.

Disturbances can determine which species live in one place and also how common each species is in that place; that is, they can have important effects on diversity. When the disturbances are small, the species that are better competitors will end up imposing themselves on the rest and dominating the community. When the level of disturbance is very high, the species that can recover quickly after each adverse event, or the best colonizers, that can quickly occupy the disturbed areas, will be the ones that dominate the community. Then, as the magnitude or frequency of perturbations changes in a particular place, one can see changes in species diversity. The intermediate disturbance hypothesis states that intermediate levels of disturbance will lead to higher species diversity values. This is because, in an ecosystem in which disturbances are rare, species best endowed for competition will eventually dominate and dominate the community. It seems that whereas in a highly disturbed ecosystem, species capable of resisting disturbances will mainly survive, in ecosystems located between these two extremes, both types of species will be present, then the diversity will be higher.

3.2.3 Ecosystem Diversity

An ecosystem is an array of microbes, animals and plant communities living in a particular habitat with distinctive physical environmental conditions and a dynamics of transfers or flows of material and energy among its components. Examples are forests, grasslands, deserts, mountains, as well as aquatic ecosystems such as rivers, lakes, and marine, coastal mangroves, of distinctive species.

Ecosystem diversity is the diversity of habitats or ecosystems within an area. A region possessing a wide variety of ecosystems will include a much greater genetic diversity, as well as species diversity, and a higher functional diversity than one in which there are only a few different habitats.

There are few areas in the world that can be considered as left untouched by human activities, so can be referred as a "natural ecosystem." Mostly natural ecosystems are found in wilderness areas. Increasingly, however, there are many kinds of damages being carried on to the natural ecosystems, degrading them and their functioning. For that reason, today, there is an increasing awareness about the importance of ecosystems diversity as it depends on the maintenance of their functions and therefore the ecological services they provide to humankind.

Ecosystem diversity has an impact on the Resilience capacity and Productivity of a Landscape where humans live. In particular, biodiversity, encompassing variation from within species to across landscapes is considered by some author as crucial for long-term resilience of ecosystem functions and their services.

3.2.4 Functional Biodiversity

Functional diversity refers to the range of processes that involve organisms in communities and ecosystems. All the biological and chemical processes of functions including flows of materials and energy (food web chains, energy pyramids, productivity), or cycles of chemical elements, water, which are required for the species to survive and to have a balance in the ecological system. Natural ecosystems important functions are for instance productivity, decomposition rates, nutrient cycling, and resistance and resilience to perturbations.

Functional diversity refers to the diversity of ecological process that maintains a balance among all the components of the ecosystems that are including all the components that are inter-dependent on each other. It comprises different kinds of ecological interactions amongst species and also ecological processes. Ecological interactions, for instance, could be defined as competitions amongst the species, parasitism or mutualism, whereas, ecological processes are nutrient retention and recycling taking place through various resources available on earth. It also includes the varying tempos and intensities of natural disturbances that many species and communities require if they are to persist. Functional diversity helps in predicting the impact of living organisms on an ecosystem and also helps in developing a connection between the two, the organism and the ecosystem.

The most difficult and critical part of developing predictive measures of functional diversity is to make a choice of various functional traits, which would be responsible in distinguishing the organisms and how the traits of those organisms would be summarized into a measure of functional diversity. Also, the functional diversity needs approval through quantitative analysis and various experiments, which is again a critical task to perform. Large amount of information is now easily available that define the traits of plant species and animal species. Choosing which traits to include in a particular measure of functional diversity will depend on the specific aims of a particular study. Quantitative methods for choosing the characteristics of a species are developed, but more efforts are required in choosing the right kind of trait. Currently, various measures are available to measure the functional diversity of the species.

Researchers mostly focus on those species, which are not supposed to be clustered in a group. Many researchers measure diversity by summarizing distances among species in trait space, whereas, there are some who focus on estimating the size of a dendrogram (grouping of species in communities) required to describe the difference, and some include information about species' abundances. There is good experimental and analytical evidence that functional diversity can provide a link between organisms and ecosystems, but greater validation of measures is required.

Every species that is an important part of the biological diversity is in some way or another dependent on each other. The service provided by one species is the life savior for another one. These services ensure the survival of each species, and therefore, it is regarded as a type of cooperation and mutual understanding for the survival of any living organism. This entire criterion of dependence amongst the living organisms is also termed as "balanced ecosystem."

The relationship between soil, plants, bacteria, and other life can be observed in the biogeochemical cycles of great importance in biodiversity as the water cycle, the phosphorous cycle, the nitrogen cycle. The nitrogen cycle can be described as the biogeochemical cycle, where the nitrogen gets converted into various chemical forms. These chemicals circulate among the atmosphere, biological system, and marine ecosystems. The nitrogen is converted through biological and physical processes. The percentage of nitrogen in the atmosphere is around 78%.

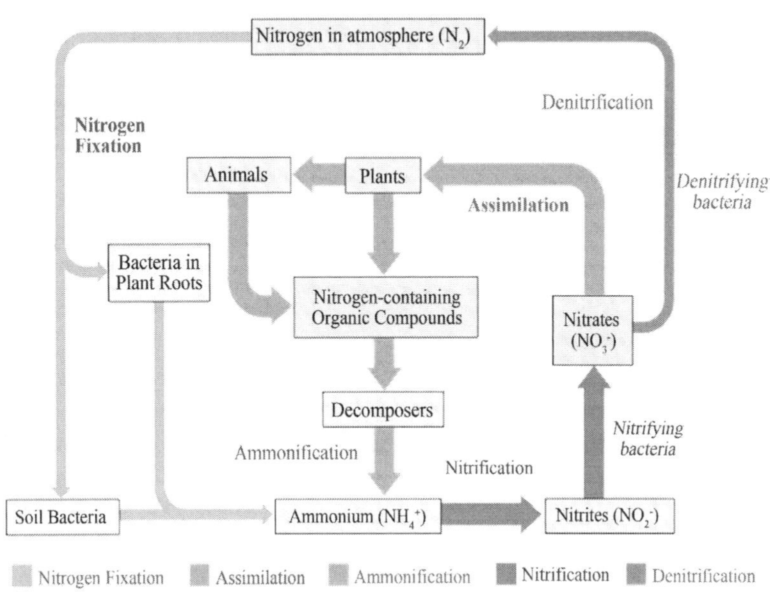

Figure 3.2: The nitrogen cycle.

Source: http://ib.bioninja.com.au/options/option-c-ecology-and-conser/c6-nitrogen-and-phosphorus/nitrogen-cycle.html

The availability of nitrogen is limited for biological purposes, which further leads to a scarcity of usable nitrogen in various kinds of ecosystems. The ecologists are always inclined towards the nitrogen cycle due to the availability of nitrogen that has a huge impact on the processes related to the ecosystem. This includes primary production and decomposition. Human activities such as fossil fuel combustion, use of artificial nitrogen fertilizers, and release of nitrogen in wastewater have dramatically altered the global nitrogen cycle.

For example, consider all the species of animals and organisms involved in a simple field used in agriculture. The activities are carried by the animals present on the field: Crop by-products feed cattle, Cattle waste feeds the soil that nourishes the crops, Crops, as well as yielding grain that also yield straw, Straw provides organic matter and fodder. Crops are therefore food sources for humans and animals, and soil organisms also benefit from crops. Bacteria feed on the cellulose fibers of straw that farmers return to the soil, Amoebas feed on bacteria making lignite fibers available for uptake by plants, Algae provide organic matter and serve as natural nitrogen fixers, Rodents that bore under the fields aerate the soil and improve its water-holding capacity, Spiders, centipedes and insects grind organic matter from the surface soil and leave behind enriched droppings. The main contribution of earthworms is generally in the form of providing fertility to the soil. They provide aeration, and drainage for soil formation and maintains the structure of the soil.

In an ecosystem, all the species are supposed to perform a particular function. These functions could be anything that would help in maintaining stability in the ecology. For example, the species could capture and store energy, produce organic resources, apart from the production many species help in decomposing the organic resources or material, helps in water purification and nutrients throughout the ecosystem, control erosion or pests, fix atmospheric gases, or help regulate climate. The ecosystem functions are transferable to ecosystem services for humans as it is impossible to survive without all these resources and services produced by the ecosystems including soil fertility, pollinators of plants, predators, decomposition of wastes, purification of the air and water, stabilization and moderation of the climate, decrease of flooding, drought and other environmental disasters.

The functioning and services of the ecosystem are influenced by the characteristics of the species available. They are not influenced by the total number of species present on this planet. The relative importance of a species to ecosystem functioning is determined by its traits and its relative abundance. For example, the traits of the dominant or most abundant plant species, such as how long they live, how big they are, how fast they assimilate carbon and nutrients, how decomposable their leaves are, or how dense their wood is, are usually the key species drivers of an ecosystem's processing of matter and energy. Thus, it is important to conserve the composition of biological communities. Increasing the number of species is not going to maintain a balance in the ecological system and its services while lowering can hamper the delicate balances.

A disproportionately huge, irreversible and negative alteration of ecosystem processes takes place when there are changes in the biotic interactions amongst the species. These interactions are done in various forms, like, predation, parasitism, competition, and facilitation. Apart from the direct interactions, there are few indirect interactions as well, such as a predator preying on a dominant competitor such that the dominant is suppressed, which permits subordinate species to coexist. Interactions with important consequences for ecosystem services include pollination; links between plants and soil communities, including mycorrhizal fungi and nitrogen-fixing microorganisms; links between plants and herbivores and seed dispersers; interactions involving organisms that modify habitat conditions (beavers that build ponds, for instance, or tussock grasses that increase fire frequency); and indirect interactions involving more than two species (such as top predators, parasites, or pathogens that control herbivores and thus avoid overgrazing of plants or algal communities).

If there is a constant shift to different rules, then a positive change could be expected in biodiversity, the services provided by the components of the ecosystem and human behavior. Alterations of ecological interactions equilibrium dynamics have been commonly documented in pelagic systems due to thresholds related to the changes constantly occurring in the climate and overexploitation of resources.

There are many regimes shifts that are essentially irreversible; for example, coral reef ecosystems that are now a part of dominated reefs (earlier were a vital part of coral dominated reefs). These shifts are mainly driven by the increasing nutrient inputs to eutrophic conditions and the removal of the herbivorous fishes, which are responsible for maintaining a balance between the corals and algae. Once the thresholds (both an upper and a lower threshold) for the two ecological processes of nutrient loading and herbivory are passed, the phase then shifts, finally resulting in a remaining less productive and diverse processes ecosystem. Consequently, human well-beings are affected not only by reductions in the food supply, but also there is a diminishing rate in the income of various human beings whose occupation is related to reefs, for example, occupations like, diving and snorkeling, aquarium fish collecting, and so forth.

Algal reefs are more prone to storm events like Tsunamis, which would further lead to shoreline erosion and seawater breaches of land. Such tremendous phase shifts are generally found in places like, Jamaica, in the Caribbean, and in the Indo-Pacific reefs. There are estimates that approximately 20% of worldwide coral reefs have already been lost by the coral-algal. Corals and algae use different physical (sweeper tentacles, mesenteries filaments, overtopping, abrasion) and chemical (allelopathy) strategies to compete for their space and maintain to preserve it. Many mechanisms of complex interactions can also involve microbes to mediate competitive interactions between coral and algae. On a reef-wide scale, it has been found that when the abundance of benthic algae is associated with an increase in microbe's density, there are stressful conditions to corals that lead to what is known as coral disease.

Food-webs equilibriums as part of the functional diversity of ecosystems can also be altered by the introduction of invasive species, which at times is considered as one of the most dramatic drivers of change in ecosystem structure, function, and delivery of services. For example, the introduction of the carnivorous bioluminescence Ctenophore *Mnemiopsis leidyi* (a jellyfish-like animal) in the Black Sea caused the loss of 26 major fisheries species and has been implicated (along with other factors) in the subsequent growth of the oxygen-deprived "dead" zone. This invasive species is considered among the list of the 100 invasive species more dangerous worldwide.

Another species among the 10 most invasive are the Asian carps introduced in North America and Europe; they are carnivorous with a high reproduction success that ends taking food and habitats from the local species until they finish eliminating them altering the local species diversity of fish and therefore the local aquatic ecosystems. A similar pattern is found in South America with the Tilapia fish species.

Eight other globally recognized invasive species in areas out of its natural range of distribution are Zebra mussels, the Cane Toad (*Rhinella marina*), European Starling (*Sturnus vulgaris*), the European/Common

Rabbit (*Oryctolagus cuniculus*), the grass Kudzu (*Pueraria montana* var. *lobata*), Asian long-horned beetle (*Anoplophora glabripennis*), Small Indian mongoose (*Herpestes auropunctatus*), Northern Pacific seastar (*Asterias amurensis*), Water hyacinth (*Eichhornia crassipes*).

Research advances indicate that "the more diverse an ecosystem the better it can withstand environmental stress and the more productive it is. The loss of a species thus decreases the ability of the system to maintain itself or to recover in case of damage." There are very complex mechanisms which focus on the ecological effects on the living organisms. The diversity of species and genes in ecological communities affects the functioning of these communities. The two main subjects, where the effect of biodiversity on ecosystem function has been studied, are the relationship between diversity and productivity, and the relationship between diversity and community stability.

3.3 PATTERNS OF BIODIVERSITY

Biodiversity is abundant and is present almost everywhere on earth and also in every drop of its bodies of water. The virtual omnipresence of life on Earth is seldom appreciated because most organisms are small (<5 centimeters); their presence is sparse, ephemeral, or cryptic, or, in the case of microbes, they are invisible to the unaided human eye.

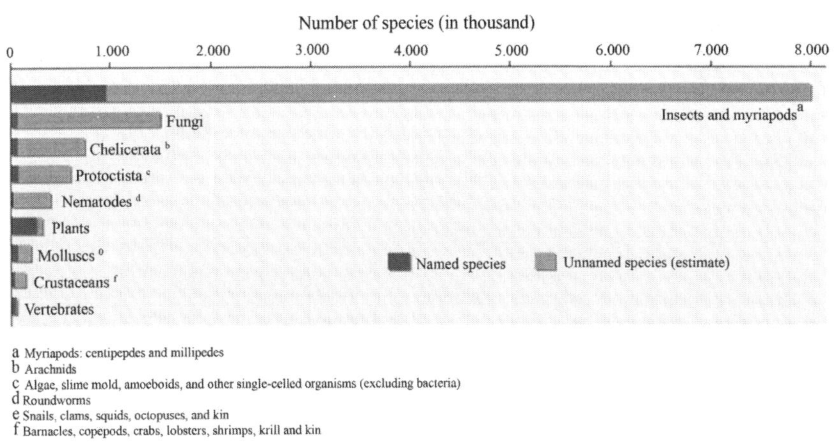

Figure 3.3: Estimates of the proportion of named species.

Source: Millennium ecosystem assessment

It is not easy to explain the spatial pattern in biodiversity due to the fact that taxonomic, functional, tropical and genetic dimensions of biodiversity need to be quantified. According to many authors, the best-known dimensions of biodiversity are not complete and are strongly biased towards the species level. This leads to a significant gap in knowledge regarding other levels such as genes and populations in plants, species, microorganisms, and so forth.

3.3.1 Spatial Patterns of Biodiversity

The spatial patterns of biodiversity comprise the distribution of hotspots, Biomes, Biogeographic Realms, Ecosystems, and Ecoregions along the earth. The spatial patterns are the global latitudinal, longitudinal and altitudinal gradients of biodiversity and the distribution of its components are modulated by the biotic and abiotic conditions of each particular region.

It is very important for conservation decisions to know the spatial patterns of biodiversity; however, it is difficult to provide a clear picture of the extent and distribution of all components of biodiversity due to the insufficiency of data within the ecologists. But still, there are many patterns and tools that decision-makers can use from existing data collections to derive useful approximations for both terrestrial and marine ecosystems. The most appropriate data on spatial distributions of species like butterflies, mammals, reptiles, plants are found in North-temperate regions and are reasonably well documented globally. Biogeographic principles (such as gradients in species richness associated with latitude, temperature, salinity, and water depth) or the use of indicators could also help to replace the need of deep biotic inventories in each of the ecosystems having similar conditions than others in close regions with similar environmental and historical factors. According to Millennium assessment reports, "Global and sub-global maps of species richness, several of which are provided in the current state and trends and scenarios, provide valuable pictures of the distribution of biodiversity."

Mostly, macroscopic organisms have comparatively small and clustered geographical ranges, which lead to high diversity and endemism. They are frequently concentrated in isolated or topographically variable regions like islands, mountains, peninsulas, and so forth. A large proportion of the world's terrestrial biodiversity at the species level is concentrated in a small part of the world, mostly in the tropics. Terrestrial vertebrates, which comprise almost one-third of all the species present in the ecosystems are located in areas with less than 1,000 square kilometers. In contrast, local and regional diversity of microorganisms are somewhere similar to large-scale and global diversity because of their large population size, greater dispersal, larger range sizes, and lower levels of regional species clustering.

Biomes and biogeographic realms give a wider picture of the distribution of functional biodiversity. Functional biodiversity illustrates the patterns of association of biota typical of wetlands, forests, grasslands, estuaries, and so forth with geography and climate known as biomes.

At lower levels, spatial patterns of species distributions in ecosystems can be conducted by analyzing frequency distributions of species abundance or dominance with modeling adjusts: to binomial if the distributions are clustered meaning in groups along the space, or Poisson if they are randomly distributed, which means that each individual of each species can be randomly found to colonize any of the available spaces.

3.3.2 Temporal Patterns of Biodiversity

Temporal Patterns of Biodiversity comprise all the background rates of extinction and biodiversity loss taken place in the ecosystems along time scales. Understanding the temporal patterns of biodiversity give the link of biodiversity to biogeography and help us to get to know more information about the species who are either in danger or who are already extinct over a geological period of time.

The Millennium Ecosystem Assessment in 2005 reported: "Except for the last 1,000 years, global biodiversity has been relatively constant over most of the human history, but the history of life is characterized by considerable change. The estimated magnitude of background rates of extinction is roughly 0.1–1.0 extinctions per million species per year. Most measurements of this rate have come from assessing the length of species' lifetimes through the fossil record: they range over 0.5–13 million years, and possibly 0.2–16 million years." These data probably underestimate background extinction rates because they are necessarily largely derived from taxa that are abundant and widespread in the fossil record.

There are few discrepancies between the dynamics of changes in natural systems and human responses to those changes. This discrepancy arises due to the time taken to respond by the ecosystems and its components, and the complexities of the feedback received from the ecologists between socioeconomic and ecological systems, and the difficulty of predicting thresholds. Multiple impacts (especially the addition of climate change to the mix of forcing functions) can cause brinks or rapid and dramatic changes in ecosystem function even though the increase in complexities and stress in the environment has been small and constant over time. Understanding such thresholds requires having long-term records, but such records are usually lacking, or the monitoring has been too infrequent, of with wrong periodicity, or too localized to provide the necessary data to analyze and predict threshold behavior.

3.4 GLOBAL PATTERNS OF BIODIVERSITY

The most prominent feature of planet earth is that it supports life. The earth constitutes of both living and non-living organism and the space occupied by such organisms is called as the biosphere. The living organism is arranged in separate groups in the biosphere; that is, there are different types of living organisms on the earth, each organism is categorized under different species, and each member of a single species differs genetically.

Table 3.1: Number of Discovered Species and their Global Total

Citation: UNEP-WCMC, adapted from tables 3.1-1 and 3.1-2 of the *Global Biodiversity Assessment*

Source: https://www.cbd.int/gbo1/chap-01.shtml

S. No.	Kingdoms	Described Species	Total Discovered Species
1	Bacteria	4000	1,000,000
2	Protists (algae, protozoa, etc.)	80,000	600,000
3	Animals	1,320,000	10,600,000
4	Fungi	70,000	1,500,000
5	Plants	270,000	300,000
TOTAL		1,744,000	Ca. 14,000,000

It is estimated that there are almost 1.75 million species that are discovered across the world. It is also believed that there are several millions of species that exist but remain undiscovered.

The above table determines the total number of species present globally. However, these estimates are expected to be incomplete because as revealed that new species are discovered every day. It is estimated that approximately

10 million species of animals are present on earth out of which only 8 million animal species, which are insects, are recognized and are formally named. In addition to this, it is also revealed that approximately 10,000 species of birds and 4,640 species of mammals are recognized. The estimated number can be inaccurate because several species are yet to bc discovered.

Apart from this, it is widely recognized that almost 71% of the surface of the earth is covered by water bodies. It is determined that these water bodies have the average depth of approximately 3.8 kilometers and the entire marine region has the capacity to support life. Thus, a vast size of biosphere comprises sea and oceans present on the planet. It is estimated that only two-thirds of the total water present on earth is frozen as ice glaciers and 2% of water is non-saline. While one-third of the world's total water is considered s groundwater that is present in the upper layer of the earth's shell.

On the other hand, the total rivers, lakes, and wetlands present in the world also known as surface freshwater hold a remaining percent of the water that is an essential sector of global biodiversity. For instance, there are 25,000 discovered species of fishes out of which 40% of the fishes produce in freshwater and large ancient lakes. A wide diversity of terrestrial ecosystem is present on land including the surface freshwater, which accounts for 29% of the earth's surface. It is projected that northern hemisphere consists most of the land surface in comparison to the southern hemisphere.

One of the significant features of biodiversity is that all the species present on earth are not evenly distributed. However, it is well known that as we move towards the equator, the richness of overall species tends to increase. This means that there is a number of species present in the tropical area in comparison to temperate regions and Polar Regions have the minimum number of species in comparison to the temperate region. The variation in the number of species is highly interrelated with variation in energy and water availability at the global level.

The value of areas, which are rich in biological diversity, is appreciated increasingly across the world and it is considered as a valuable asset. Therefore, World Heritage Convention, which is an International agreement organization, attempts to preserve and conserve such areas that have rich biodiversity.

The biodiversity of the planet is spread in few specific ecological areas, and there are thousands of ecoregions across the world. Out of the total number of Eco regions in the world, 200 regions are considered as the riches, distinctive and rare Eco regions. These natural regions are also known as

"Global 200." In addition to this, it is projected that 50,000 plants that grow in a certain region comprise 20% of global plant life. This is because these endemic plants produce only in 18 hotspots across the world.

Nations that have a large number of hotspots of diversity are called as "mega diversity countries." The extinction rate of species across the world is ambiguous, but it is determined that rate of extinction is extremely high because wilderness regions are decreasing at a fast rate. Moreover, the diversity hotspots, which are globally accepted, are mostly situated in Western and North-East Ghats of India these areas are considered as most biodiversity-rich areas in the world.

Table 3.3: Number of Total living species that are Currently Known

Source: https://www.ugc.ac.in/oldpdf/modelcurriculum/Chapter4.pdf

S. No.	Group	Number of Described Species
1	Bacteria and blue-green algae	4,760
2	Fungi	46,983
3	Algae	26,900
4	Bryophytes (Mosses and Liverworts)	17,000
5	Gymnosperms (Conifers)	750
6 Angiosperms (Flowering Plants)		250,000
7	Protozoans	30,800
8	Sponges	5,000
9	Corals and Jellyfish	9,000
10	Roundworms and earthworms	24,000
11	Crustaceans	38,000
12	Insects	751,000
13	Other Arthropods and minor invertebrates	132,461
14	Mollusks	50,000

15	Starfish	6,100
16	Fishes (teleosts)	19,056
17	Amphibians	4,184
18	Reptiles	6,300
19	Birds	9,198
20	Mammals	4,170

It is noteworthy that human activities have caused great loss to biodiversity including habitat destruction, extinction of species, disruption in ecosystem process and also eroded genetic diversity within species. For understanding and conserving the biodiversity, it is essential that it is measured in an effective manner by measuring the richness of species. As determined millions of species are identified already, but there are many species yet to be discovered. Discovering, naming and investigating the relationship between the species is based on highly skilled taxonomists. Species rate of extinction has increased 1,000 times in comparison to the past due to human activities. According to The Millennium Ecosystem Assessment (MEA) report 2005, 10–30% of the amphibian species, bird species, and mammal species are currently under the threat of extinction.

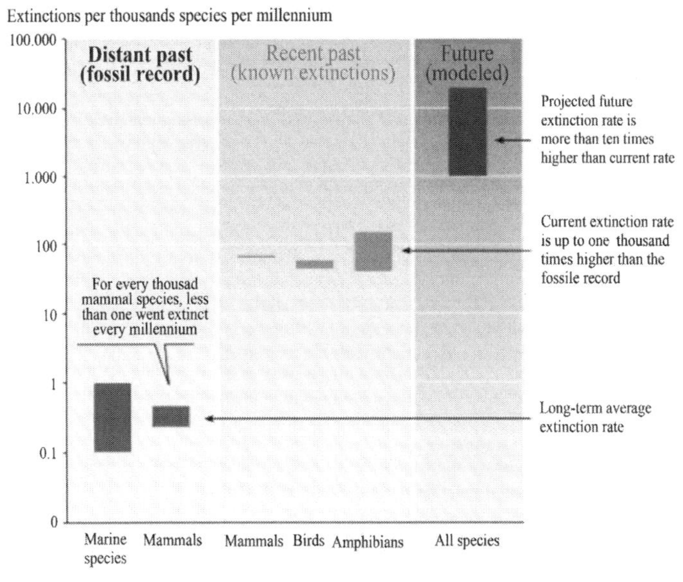

Figure 3.4: Past and Future Number of Species Extinction. Source:

It is also estimated that the current rate of extinction of species is unsustainable. It is revealed in a study that almost 20 mammal species extinct during the last century, which is a great loss to the planet. The human beings are overusing or misusing the natural ecosystem due to which productive forests, grasslands have all declined to a great extent. This caused the destruction to wilderness areas and habitat, which further resulted in the loss of species due to the destruction of the natural ecosystem.

3.4.1 Global Trends at Genetic Level

The difference among genes is fundamental to genetic diversity that is the difference between the functional units of heredity. However, it should be noted that all the genetic material does not have a role in heredity, but in few organisms a large part of DNA is repetitive. Genetic diversity is essential because it allows adaptation and transformation in organisms. When there is any type of environmental change, the organisms with a high level of genetic diversity are capable of facing such changes. In such phase, organisms with low genetic diversification are at increased risk of extinction. Change in genetic diversity of organisms is highly due to increasing human activities that affect biodiversity.

Genetic diversity or genetic variation is a term that refers to the idea that different members in a population carry different genetic sequences. For instance, the genetic sequence in human beings is also different such that one individual may carry a genetic sequence that code his blood group as "A" while the other individual may carry a different sequence that code for type "O" blood. In a similar manner, few individuals may carry different genetic sequence due to which they do not code for anything at all.

Genetic diversity is also decreased to zero in few extreme cases when the species are near to extinction. When human activities result in loss of genetic diversity, it is called as genetic erosion. For example, it is estimated that black rhinos are heading towards extinction from Africa because of human activities. The horns of rhinos are in huge demand among the population for production of the dagger, which is used, in a ceremonial event for Middle Eastern puberty rites. Other than this, elephants, cats, and whales are among other species, which are threatened.

According to a research conducted by Cardiff University, an international research team compared the genes of all the existing and extinct rhinos for the first time, where it was found that there was a huge decline in genetic diversity of rhinos. It was found that out of 64 genetic families of rhinos

44 genetically diverse species of rhinos exists no longer. It is also revealed that in last two centuries the evolutionary capabilities of black rhino have declined drastically due to hunting and huge loss of habitat. The human activities have caused a huge decline in the genetically diversified species, and they are incapable to adapt to the climatic changes that cause their extinction. The landscape change and ecosystem are threatened due to increased pressure from humans that several genetically diverse species are endangered.

Genetic diversity is an essential element that plays a major role in in the overall health of the species and their ability to fight against diseases. For example, a mass extinction event occurred about 12,000 years ago, and it is believed that events eliminated about 75% of the large species of mammal worldwide. However, few species such as Cheetahs managed to survive, but that event led to the loss of their genetic diversity along with inbreeding. As a consequence, the Cheetahs that exist today have poor genetic health and sperm quality, and, along with this, they are prone to diseases.

Figure 3.5: Distribution of Cheetahs Sub-species.

Note: Light colors denote historic range before 1900.

Source: https://cheetah.org/about-the-cheetah/genetic-diversity/

Let's take another example or a case of chipmunks. The researchers collected DNA of chipmunks from museum samples population of chipmunks in the early 1990s and compared those DNA samples with the DNA samples of chipmunks collected in last 10 years. It was found that the current population of the chipmunks is found to be more similar genetically. That is, the present population of chipmunks has a fewer difference in genetic diversity in comparison to their early counterparts. However, the researchers conducted the research in few non-coding locations it is estimated that all the chipmunk's genome have lost genetic diversity due to climatic change.

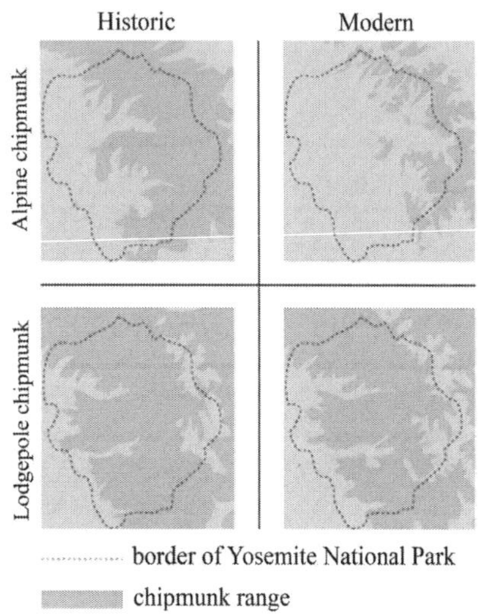

Figure 3.6: Population size of alpine chipmunk and climate change? Source?

As shown in the above figure, the population size of alpine chipmunk has declined significantly in last 90 years due to climate change. The climate has become warmer at present in comparison to a century ago. And it is considered that due to decline in population size of the chipmunk, there is also a loss in genetic diversity of chipmunk. For instance, few species of chipmunk such as lodgepole chipmunk also biologically named, *T. speciosus* is not as affected as much as alpine chipmunks.

It is found that there is a similar level of genetic diversity in the present and early samples of lodgepole chipmunk as they are lower-elevation species. On the other hand, change in temperature has affected alpine chipmunk in a negative manner, as these genetically different species of a chipmunk is a high-elevation species. The main difference between the two genetically different species is their response to increasing temperature. The change in climatic conditions is the major cause of declining genetic diversity in the alpine chipmunk.

If the genetic diversity continues to decline, there will be several consequences for alpine chipmunks such as:

- Loss of genetic diversity raise the idea that species population is becoming more fragmented due to that is, the subpopulation of the species that are genetically diversified are being separated from one another, which increases the probability of species extinction.
- It is considered that genetic diversity is the core element of development, therefore, if the genetic diversity if affected due to change in environment, the ability of the species to evolve will also reduce. It is noteworthy that if the species do not have the right gene that helps them in adaptation to the environment, they will be endangered.
- Another consequence of reduced genetic variation is that it can lead to a decreased rate of reproduction and survival through inbreeding depression. For instance, when there is a lower level of genetical difference in the members of a species, there can be a situation where the two genetically different members who are close relatives might mate with each other. This situation may lead to reproduction of an offspring with reduced fitness.

All the three factors mentioned above have the possibility that rate of reproduction of chipmunks might decline to a great extent that might further result in their extinction. Therefore, it is considered that genetic diversity should be present at a suitable level so that the species population has the ability to adapt to the environmental changes and overcome unexpected disasters.

3.4.2 Global Trends at Species Level

It is considered that the most important component of loss of biodiversity is the extinction of species. Extinction and decline of species is a natural part of evolution, but the current pattern of extinction is not natural because the current extinction rate of species is recorded as much higher in comparison to historical records. The present status of species extinction is recorded that at least 1,000 species are lost every year. That accounts for an increased extinction rate that is 1,000 times higher than the natural rate.

The fossil record reveals that all the species have a specific life-span and they die after that. It is also recorded that a large proportion of species that have ever existed are now extinct. The extinction occurs when the species die without producing any offspring. There is two major cause of species extinction:

- The deterministic process is also known as cause and effect relationship. For example, species extinct due to direct human involvements such as conversion of land and deforestation.
- Another cause is the stochastic process that can occur by random events such as demographic uncertainties, environmental uncertainties, and catastrophic events.

Few scholars projected that about 25% of the species of mammals might extinct in the near future. It is also estimated that 37% of the hoofed mammals, 50% of the primates (including monkeys, apes, and lemurs) and 26% of the bats and meat-eaters are at high risk of extinction. Marine species such as fishes are not an exception. It has been evaluated that 30% of the fish species are endangered along with 20% reptiles and 25% amphibians among other threatened species.

According to a report of World Wildlife Fund released in the year 2016, the population of the species living in freshwater has declined by 81% between 1970 and 2012. Another report released by Living Planet (2016), the species of freshwater decline is more than the decline of species that is observed on land and marine water. It is estimated that 38% of the land species and 36% of the marine species are under declining phase. The major reason behind declining species population living in freshwater is the loss of habitation. The lakes, wetlands, and rivers around the world are continuously getting polluted, fragmented, damaged and abstracted due to which diversity of freshwater species is declining. Several types of research revealed that human activities are the major cause of freshwater habitation loss. Another cause of loss of freshwater species is over-exploitation of natural resources. Humans harvest birds and fishes for satisfying hunger, and other species such as reptiles and amphibians are caught for the purpose of the pet trade. It is also reported that in 2014 the species population of freshwater has declined by four-fifth.

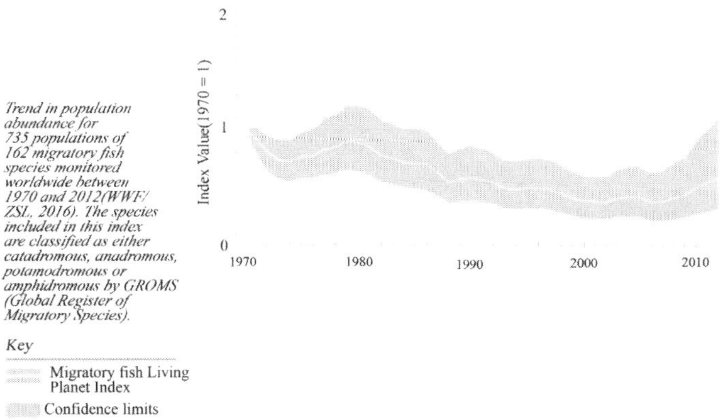

Figure 3.7: The migratory fish LPI shows a decline of 41% (range: +12 to −69%) between 1970 and 2012.

Source: https://freshwaterblog.net/2016/10/27/freshwater-species-populations-fall-by-81-between-1970-and-2012/

The above figure depicts that between the year 1970 and 2012, there was a drastic decline in migratory fish species. The river fragmentation is the major cause of the decline in migratory fish species as fragmentation likely blocks the natural migration route, which ultimately impacts their producing success and health. In addition to this, it is also determined that between 1970 and 2012, the global population of the species of amphibians, birds, reptiles, fishes, and mammals decreased radically by approximately 58%. The current trend of loss of species in the world is taking place at the rate of 2% per year, which is estimated to remain constant in the future.

According to Whigham (2016), the total animal population loss across the world is approximately 60% since 1970, and the estimations of the researchers depict that these trends will get worse in the near future. The human population is largely dependent on animals for food as we continue to swallow up the large population of animals; the biodiversity of the earth is affected to a great extent because the animal population is declining at an alarming rate. Human activities are affecting biodiversity, and thus, the world is heading towards mass extinction. A large number of animal species has already evaporated in front of the human population.

The number of vertebrate animals' species has also declined in plenty between 1970 and 2012 by 58%. It is estimated that if the trend of species loss continues, approximately two-thirds of all species of amphibians, birds, mammals, reptiles, and fishes will disappear in the near future. The major factors that contribute to the declining biodiversity of the planet are considered to be habitat loss and its degradation along with unsustainable use of natural resources and climatic change.

3.5 CONSERVATION OF BIODIVERSITY

Conservation of biological diversity is very essential to conserve ecological diversity and to preserve the continuity of food chains. Conservation of genetic diversity leads to preserve plants and animals' lives and therefore the humankind as users of that genetic diversity. This is to ensure the sustainability of life support system on earth. A vast section of knowledge for the potential use of natural resources is provided by to the scientific community. Through conservation of biodiversity a reservoir of wild animals and plants are preserved. Biological diversity provides immediate benefits to the society such as recreation and tourism. Biodiversity conservation serves as an insurance policy for the future. Biodiversity and its conservation is very essential for the survival of human race. The conservation of biodiversity can be achieved following two different strategies: In-situ conservation and Ex-situ conservation.

3.5.1 In-situ Conservation

The on-site conservation is also known as in-situ conservation, which is generally required for the conservation of the genetic ecosystems and their resources. Biodiversity at all its levels, genetic, species and intact ecosystems, can be best preserved in-situ by keeping a suitable representation of wilderness as protected areas. For in-situ conservation of biodiversity, a network of national parks and wildlife sanctuaries are kept into consideration for local to national governments and regional or global organizations. Such a network would preserve the total diversity of life of a region. In many countries, National parks and Wildlife sanctuaries were regarded as a preserving area for wildlife species, the main aim of these parks and sanctuaries is to preserve the natural ecosystem in an intact manner where biological diversity—from microscopic unicellular plants and animals, to the giant trees and major mammals—can all be preserved.

The in-situ conservation of biodiversity is the process of protecting those plants and animal species that are at an edge of getting extinct in its natural habitat. This process is carried forward by protecting and cleaning the habitat and also by defending the species from the hunters. However, it is not an easy task for an individual to preserve the species alone because all the species in an ecosystem are inter-dependent on each other. Thus, the whole ecosystem must be protected. According to biologists, it is advisable to do in-situ conservation in those areas that are relatively rich in species or areas where the endangered species are found. The methods adopted for in-situ conservation are:

- Biosphere reserves: there are around 18 biosphere reserves in India that are helping in conserving biodiversity.
- National parks: there are around 104 national parks in India for the same purpose.
- Wildlife sanctuaries: around 537 wildlife sanctuaries are constructed and is one of the most effective ways to conserve biodiversity.
- Biodiversity hotspots:

In small and remote areas endemic species are generally found. Many ecologists believe that these areas should be given special treatment so that these endemic species do not get extinct by human activities. Animals such as elephants require different types of habitat to feed in during different seasons. A vast area is required so that they could utilize them to have highly nutritious products for their sustenance. When the grasslands get dry, elephants move back to the woods to feed on foliage from the trees. "A Protected Area that is meant to protect elephants must, therefore, be large enough and include diverse habitat types to support a complete complement of inter linked species." Jaguars in America require extensive rainforests corridors and large bird species like the Harpy Eagle whose habitat is the trees canopies of rain forests.

The International Union for Conservation of Nature and Natural Resources states that "it is essential to include at least 10% of all ecosystems as Protected Areas if biodiversity is to be conserved in the long-term.

Human activities are speeding up in such a way that it is now understood and also visible at the same time that very few good grasslands are available in all around the globe and are regarded as protected areas. There are grasslands that have been overgrazed and are now popularly known as wastelands. Such wastelands have a low biological value, and they demand improved and careful management system to get back into a more "natural state." Only a few wetlands have been made into Sanctuaries. These require better management. The main advantage of having in-situ conservation of biodiversity is that it helps in recovering populations in the environment where the distinctive characteristics of resources have been developed. The in-situ conservation practices ensure that there is an ongoing process of evolution and adaptation of living organisms and plant species within a given framework.

3.5.2 Ex-situ Conservation

The conservation of biodiversity and its components that are present outside its natural habitats are determined as Ex-situ conservation. This involves conservation of genetic resources, as well as wild and cultivated species, and draws on a diverse body of techniques and facilities. Some of them are:

- Gene banks, for example, seed banks, sperm and ova banks, field banks;
- In vitro plant tissue and microbial culture collections;
- Captive breeding of animals and artificial propagation of plants, with possible reintroduction into the wild; and
- Collection of living organisms for various zoos, botanical gardens, so that ecologists and biologists are able to conduct research and public awareness on those particular species.

Ex-situ conservation measures are balancing in nature as they carry forward the methodologies adopted by the in-situ conservation practices. This is also a complementary activity as it provides an insurance policy against the extinction of the species. The ex-situ conservation practices also play a crucial role in recovering and initializing programs for the species who are regarded as endangered species. "The Kew Seed Bank in England has 1.5% of the world's flora—about 4,000 species—on deposit."

Apart from conserving the extinct species, ex-situ conservation plays a vital role in the field of agriculture as well. In agriculture, ex-situ conservation measures maintain domesticated plants, which are not able to survive in nature unaided atmosphere. This conservation is also considered as one of the best research opportunity for biologists to perform research on biological components of biodiversity. Some of these institutions also play a central role in public education and raising awareness by bringing members of the public into contact with plants and animals they may not normally come in contact with. "It is estimated that worldwide, over 600 million people visit zoos every year." Since both ex-situ and in-situ conservation practices are complementary to each other, it is important to provide support to both these measures although the primary objective for conserving biodiversity should be in-situ conservation. Conservation of a species could be done in a better manner only if it helps in protecting the habitat of all the species that are present on this planet. When there is a species, which is on the edge of getting extinct, then alternate methods are adopted to carefully control the situation. Construction of a botanical garden is one way through which the species of plants and their sustenance could be tackled intelligently. According to biologists, "These breeding programs for rare plants and animals are however more expensive than managing a Protected Area. There is also another form of preserving a plant by preserving its germplasm in a gene bank so that it can be used if needed in future. This is even more expensive."

Breeding is another method to help the endangered species of animals. However, the most important step of a successful breeding program is the reintroduction of a species into its original wild habitat. This requires rehabilitation of the degraded habitat and removal of the other causes such as poaching, disturbance, or other manmade influences that have been the primary cause of reducing the population of the species. Earlier, domestic animals were first selected and then were bred for their ability to adapt to local conditions. Traditional agro pastoralists in India, for instance, have selectively bred livestock for 2000–3000 years. "India has 27 breeds of cattle, 40 breeds of sheep, 22 breeds of goats, and 8 breeds of buffaloes. These traditional breeds must be maintained for their genetic variability."

Local crops also have been domesticated, and local farmers and indigenous communities in America and Africa keep thousands of varieties of cereals, tuberous plants of high interest in food security.

REFERENCES

1. Barott, K., Smith, J., Dinsdale, E., Hatay, M., Sandin, S., & Rohwer, F. (2009). Hyperspectral and physiological analyses of coral-algal interactions. *PLOS One*, *4*(11), e8043. Retrieved from http://doi.org/10.1371/journal.pone.0008043
2. Cantarello, E., Newton, A. C., Martin, P. A., Evans, P. M., Gosal, A., Lucash, M. S. (2017). Quantifying resilience of multiple ecosystem services and biodiversity in a temperate forest landscape. *Ecology and Evolution*, *7*(22), 9661-9675. Retrieved from https://doi.org/10.1002/ece3.3491
3. Chivian, E., & Bernstein A. (Eds.). (2008). *Sustaining life: How human health depends on biodiversity* (illus. ed.). New York: Oxford University Press. https://phys.org/news/2017-08-biodiversity-earth.html#jCp
4. Gaston, K. J., & Petchey, O. L. (2006). Functional diversity: back to basics and looking forward. *Ecology Letters, 9*(6), 741-758. Retrieved October 13, 2017, from https://www.ncbi.nlm.nih.gov/pubmed/16706917
5. Greenfacts.org. (2005). Biodiversity: *What is it, where is it, and why is it important?* Retrieved October 13, 2017, from https://www.greenfacts.org/en/biodiversity/l-3/1-define-biodiversity.htm
6. Larsen, B. B., Rhodes, M., & Wiens, J. (2017). Inordinate fondness multiplied and redistributed: the number of species on earth and the new pie of life. *The Quarterly Review of Biology, 92*(3), 229-265. doi: 10.1086/693564.
7. Magurran, A., & McGill, B. J. (Eds.). (2011). Biological diversity: Frontiers in measurement and assessment (1st ed.). Oxford: Oxford University Press.
8. Shah, A. (2014). *Why is biodiversity important? Who cares?—Global issues*. Retrieved October 13, 2017, from http://www.globalissues.org/article/170/why-is-biodiversity-important-who-cares
9. Slingenberg, A., Turner, K., Braat, L., Windt, H., Rademaekers, K., & Eichler, L. (2009). *Study on understanding the causes of biodiversity loss and the policy assessment framework*. Retrieved October 13, 2017, from http://ec.europa.eu/environment/enveco/biodiversity/pdf/causes_biodiv_loss.pdf

10. Who.int. (2007). *WHO|Biodiversity*. Retrieved October 13, 2017, from http://www.who.int/globalchange/ecosystems/biodiversity/en/
11. Wilson, E. O. (1992). The diversity of life: With a new preface (Questions of science) (2nd ed.). New York: W. Norton & Company.

4
CHAPTER

BIODIVERSITY AND ECOSYSTEM SERVICES

"Lack of accountability weakens the environmental and health rights of citizens; it damages peace-building and reconciliation initiatives; impedes the implementation of global health policies; leads to the loss of ecosystems and biodiversity; and weakens democracy, justice, human rights, and international security."

—Widad Akrawi

CONTENTS

ABSTRACT

Biodiversity and ecosystems are gaining huge attention across the world. The loss of biodiversity and its ecosystem services has a threatening effect on the human lives; therefore, this topic is considered as the concerning issue. In particular, because one of the most important services is the regulation of global warming and the effects of climate change, thus, this chapter elaborates more on the direct and underlying causes of biodiversity loss. It also evaluates how the biodiversity loss impacts the lives of human beings. Furthermore, the role of biodiversity in ecosystem stability and conservation is also discussed.

4.1 INTRODUCTION

The wellbeing of humans is dependent on the nature that provides a variety of advantages such as food, clean water, carbon, soil, medicines, tree shade, fibers, charcoal, among many others. Without these ecosystem services present today in the natural flow in a continuous manner, the human society cannot be able to keep flourishing. Biodiversity is a glorious contribution of nature on earth. Ecosystem services provide humans food and water and other goods without charging any cost. Therefore, these are rarely identified by our present economic scope. Therefore, there has been a noticeable decline in biodiversity as the ecosystems of the earth are continuously destroyed.

On the other hand, biodiversity is essential to maintain different services of the ecosystems such as the production of food, regulation of atmosphere's chemical composition, the supply of raw materials, recycling of nutrients, provision of water and natural control of flora and fauna's population. Biodiversity is also essential for services such as genetic resource usage and various leisure activities, but nowadays it is evident that the world has already lost a considerable amount of its biodiversity. This loss of biodiversity has an adverse effect on the lives of human beings but also in affecting economies of the states as it imposes a pressure on the prices of food and commodities. For this reason, humans now recognize the need for urgent remedial actions to be taken because the loss of biodiversity is inseparably associated with the wellbeing of humans

Traditionally, it is believed that the role of nature is as the nurturer if the human beings. However, in the present environmental context, Mother Nature is increasingly at a loss with declining number of species and worsening environment. Many species such as tigers, rhinos, pandas, elephants are facing extinction. Other than this, natural disasters are increasingly taking places such as droughts, earthquakes, floods, and landslides. Along with this, the scarcity of water and food have been noticed, which has grabbed the attention of the world.

There is an instant need for evaluating biodiversity and its evolution in an effective manner; it also means conserving biodiversity to save ourselves. Environment disruption is the major reason that causes changes in variety and composition of plants, animals, and various other species. It is also responsible for an alteration in patterns of plants and animals, which are highly recognized by people living in the rural areas. The increasing greed of human beings is satisfied through evolutionary progression, which accelerates the extinction of natural resources as well as harm biodiversity. The scarcity of biodiversity seriously impacts the entire living beings on the planet.

Biodiversity is greatly threatened, and many species on earth are at the edge of extinction, in particular, those species more valuable to humans and therefore more intensively used. This is because millions of trees are cut down every day due to which forests are cleared and thus, there is no food and water for the living organisms over there. Given the scarcity of food, water and air many organisms die, and therefore, the rate of extinction is increasing. According to international NGO's the tropical forests are reducing at the rate of 0.8% every year. At this rate of extinction of plants and other living creatures, the survival of human being is also threatened. There are few international organizations that work for maintaining the biodiversity: International Union of Conservation of Nature (IUCN), Conservation International (CI), World Wildlife Foundation (WWF),

If the businesses continue to work as usual, there will be serious consequences on biodiversity and ecosystems by the year 2050. It is estimated that 11% of the natural areas will be lost by 2050 majorly as a result of a change in climate, infrastructure expansion, and agricultural conversion. It is also forecasted that almost 40% of the agricultural land that is presently under low-impact agriculture form can be converted to the exhaustive usage of agriculture that will lead to further loss of biodiversity. By the year 2050, 60% of coral reefs are estimated to be lost through different human activities such as fishing, pollution, climatic change, diseases, and coral bleaching.

Furthermore, the current land and ocean trends reveal that biodiversity is severely at threat and it represents a danger to health and welfare of human beings. Changing climatic conditions are further worsening the problem of biodiversity loss. The poorest human populations and countries across the world are at the large risk due to climate change because they are the ones who are the most and foremost influenced. This is because poor people are the ones who are most reliant on the services of the ecosystems.

The productivity of an ecosystem is dependent upon the boost of its complex matter and energy transfer interactions among populations of different species in several plant, animal and microbial communities, and all the species (whether big or small) contribute in balancing the ecosystem.

4.2 ECOSYSTEM SERVICES

According to one of the most widely cited definition in the Millennium Ecosystem Assessment, ecosystem services are described as "the benefits that people obtain from ecosystems." Fisher in the year 2009 proposed a definition in context to economic valuation and provided a distinction between ecosystem services and its benefits to the human society. He stated that "ecosystem services are the aspects of ecosystems utilized (actively or passively) to produce human well-being. Ecosystem services as being the link between ecosystems and things that humans benefit from, not the benefits themselves." Ecosystem services include ecosystem organization or structure (the ecosystem classes) as well as ecosystem processes and functions (the way in which the ecosystem operates). The processes and functions churn out to be serviced when they are taken care by the humans and are beneficial to them. In other words, "ecosystem services are the ecological phenomena, and the benefit is the realization of the direct impact on human welfare."

Biodiversity plays an important role in ecosystem services as it provides support, provision and cultural services to all the living organisms that are the most important components of biodiversity. For the sustenance of human life, these services are highly important. A great effort has been made to provide and illustrate a relationship between biodiversity, regulating and supporting services, and human well-being.

The ecosystem services are also affected by the reduction of population and extinction of species as they are no longer contributing towards the ecosystem functioning. Local extinctions (the loss of a species from a local area) and functional extinctions (the reduction of a species such that it no longer plays a significant role in ecosystem function) are taken more into consideration rather than global extinction. Global extinction is the loss of all individuals of a species from its entire range. Loss of ecosystem functions, and the services derived from them, however, occurs long before global extinction. Often, when the functioning of a local ecosystem has been pushed beyond a certain limit by direct or indirect biodiversity alterations, the ecosystem-service losses may persist for a very long time.

If any of these services get replaced or misused by any living organism, then it leads to a big disaster. Hence, it is important to conserve and sustain the services provided by the biological diversity to enjoy all the above-mentioned services in a healthier manner. According to a report flashed by Nature Magazine, "genetic diversity helps to prevent the chances of extinction in the wild (and claims to have shown proof of this). To prevent the well-known and well-documented problems of genetic defects caused by in-breeding, species need a variety of genes to ensure successful survival. Without this, the chances of extinction increases."

According to the Millennium Ecosystem Assessment definition, the ecosystem services are classified into four services. Those four services are:

4.2.1 Provision Services

Those that describe materials and energy outputs from ecosystems

- Food: Biodiversity provides the most important component to sustain life, and that is food. The food from various forms, like, crops, livestock, and fish. Biodiversity is a very important factor for the economic sustenance of human life. Biodiversity is regarded as the first source and resource for daily life. Biodiversity is regarded as the first source and resource for daily life. According to most of the people, biodiversity is a reservoir of resources that could be withdrawn for the manufacturing and production of food, cosmetics, furniture, and other cosmetic products. Thus, it is visible that shortage of resources is directly related to the loss of biodiversity and gain for the living organisms.
- Raw material: materials for construction like timber, fibers, among others. Fibers for clothing, wood for shelter and warmth, all these are the economic commodities produced and manufactured in industry. Biodiversity is also a source of energy, that is, biomass. There are many other industrial products that are produced through the resources and services provided by the ecosystem. Those industrial products are, oils, lubricants, perfumes, fragrances, dyes, paper, waxes, rubber, latexes, resins, poisons, and cork can

all be derived from various plant species. Supplies from animal origin are wool, silk, fur, leather, lubricants, waxes. Animals may also be used as a mode of transportation.

- Water: it is a substantial part of all living beings being vegetation and in particular the forest and peatland ecosystems being those with the highest importance in water provision.
- Medicinal resources: Various species of plants hold medicinal components and are also considered as pharmaceutical drugs that act as a life savior. All medicinal resources and drugs are important biological resources that help the living organisms sustain a healthy life. Wild plant species have been used for medicinal purposes since before the beginning of recorded history in all cultures along the world. For example, quinine comes from the bark of the Cinchona tree that since times of Humboldt was told about the importance to treat malaria; digitalis from the Foxglove plant (chronic heart trouble), and morphine from the Poppy plant (pain relief). According to the National Cancer Institute of the United States, "over 70% of the promising anti-cancer drugs come from plants in the tropical rainforests. Animal may also play a role, in particular in research. It is estimated that of the 250,000 known plant species, only 5,000 have been researched for possible medical applications."
- Ornamental plants, ornamental fish

4.2.2 Regulating Services

Those benefits obtained from the regulation of ecosystem processes, like:

- Carbon sequestration and storage. Is the amount of carbon that is captured from CO_2 of the atmosphere through the process of photosynthesis and accumulated in the biomass of the producers in terrestrial and aquatic (marine and freshwater) ecosystems?
- Local climate and air quality. As ecosystems influence the local climate and air quality through the biogeochemical cycles purifying the atmosphere, dissipation of thermic energy
- Biological control. Biological interactions can help in providing mechanisms to control pest and diseases of crops and livestock.
- Erosion prevention and maintenance of soil fertility. Vegetation cover and complex vegetation structures are very important to

sustain those regulating services to avoid lad degradation.

- Pollination. Interbreeding can only be possible through the participation of pollinators allowing sexual reproduction to be possible in plant species. It maintains genetic variability among crops populations increasing resilience against effects of climate change.
- Waste-water treatment. Wetlands potential for waste-water treatment is originated by all organisms that intervene in processes of the biogeochemical cycles. For instance, in wetland habitats, the ecological processes of Carbon Cycle, Nitrogen cycle and Phosphorous cycles are mediated by microorganisms and invertebrates actively working in transforming the original materials mediated with a cleaning effect if managed.
- Moderation of extreme events. Complex vegetation biomass in local communities can be very important in moderating extreme events such as storms and flooding thought the participation of ecohydrological processes.

4.2.3 Habitat or Supporting Services

Those services provided by the ecosystems to living organisms to sustain a healthy life.

- Nutrient storage and recycling: maintain the nutrients available in the ecosystem helping the living organisms to grow and develop sustainably.
- Pollution breakdown and absorption: The natural ecosystem helps in breaking down pollution. Afforestation can help in absorption of CO_2 from the atmosphere.
- Contribution to climate stability. Absorption of CO_2 from the atmosphere can help in mitigating global warming contributing to climate stability.
- Maintenance of ecosystems. Support habitats for different species.
- Recovery from unpredictable events. The maintenance of healthy ecosystems provides resilience to disturbances and protection of extreme, unpredictable event.

4.2.4 Cultural Services

- Esthetic: Cultural ecosystem services have been included in many other typologies of ecosystem services and referred as cultural services, life-fulfilling functions, information functions, amenities and fulfillment, cultural and amenity services, or sociocultural fulfillment.

- Recreational: Biodiversity is a source of economic wealth for many areas, for example, famous parks and forests where wild nature and animals are a source of beauty and joy for tourists arriving there.
- But also, spiritual and other non-material benefits.

4.3 BIODIVERSITY LOSS AND ITS CAUSES

The loss of biodiversity is the loss of living organisms and the services provided by the ecosystems. The most frequent and direct causes leading to the loss of biodiversity are the degradation and loss of habitats, exploitation and misuse of natural resources, various kinds of pollution like air, water, noise land, and so forth, and climatic changes leading to natural disasters. These factors being undistinguishably linked with some or all of the other direct causes are in turn driven by underlying causes. Generally, biodiversity loss differs from place to place and people or human species responsible for all the above mentioned-causes.

4.3.1 Loss and Destruction of Habitat

This is one of the greatest threats to biodiversity. Habitat loss is directly linked to human induced pressures on land. Alterations in Ecosystem Composition Assemblages of species and their interactions with their ecosystems are critical for not only saving the species but also for their successful future evolution. In the event of evolutions, either within species groups, or within the environment, entire ecosystems can begin to change. Alterations to ecosystems are critical factors contributing to species and habitat loss.

4.3.2 Invasive Alien Species

The introduction of exotic species that replace local and native species is cited as the second largest cause of biodiversity loss. Alien-invasive species replace and often result in the extinction of native species. The annual economic damage caused by invasive plant and animal species is estimated to be in the region of US$ 1.4 trillion.

4.3.3 Over-exploitation

Over-exploitation or unsustainable use of natural resources is another major cause of the biodiversity loss. There is a limit for every natural resource available on earth. The excessive use of these resources leads to the degradation of the environment. For example, shrimp farming in India, Thailand, Ecuador and Indonesia resulted in Wetland destruction, pollution of coastal waters and degradation of coastal fisheries. According to various scientific studies, "cost of environmental degradation resulting from shrimp farming was costing more than the earning through shrimp exports." It is the same case of Cod fish in the Baltic Sea or Characids in the Amazonia and Orinoquia fisheries. Such over exploitation of natural resources makes them a scarce resource. Some of the major consequences of the unsustainable use of natural resources are deforestation, soil erosion, the occurrence of a natural calamity, ozone air depletion, and so forth. The unsustainable use of natural resources is considered to raise the economic growth of a particular country. But, that is not true. It is a big loss for living organisms and biodiversity. Over-hunting, over-fishing or over-collecting of a species can quickly lead to decline. Changing consumption patterns of humans is often cited as the key reason for this unsustainable exploitation of natural resources.

4.3.4 Pollution and Contamination

Biological systems respond slowly to changes in their surrounding environment. Pollution and contamination cause irreversible damage to species. Pollution is considered as a major threat to biodiversity. It is not possible to overcome this threat easily. The pollution caused is of various kinds: Water pollution, noise pollution, and air pollution as some of the major kinds of pollution experienced globally. Pollutants (of any kind) are not bounded by boundaries. For example, agricultural run-off, which contains a variety of fertilizers and pesticides, may seep into ground water and rivers before ending up in the ocean. Atmospheric pollutants drift with prevailing air currents and are deposited far from their original source.

In addition, water pollution is especially injurious to the aquatic animals and coastal ecosystems. Toxic wastes entering the water bodies disturb the food chain and so the aquatic ecosystems. Moreover, insecticides, pesticides, Sulphur and nitrogen oxides, acid rain, ozone depletion and global warming too, are some of the major side-effects of pollution prevailing in biodiversity and are affecting plants and animal species in an adverse manner. All the coral reefs are getting adversely affected due to the pollutants coming out of industries, oil transportation, and offshore mining. Then comes noise pollution, which is a major reason behind the extinction of wildlife. According to the Canadian Wildlife Protection Fund, "Arctic Whales are seen on the verge of extinction as a result of increasing noise of ships, particularly ice-breakers and tankers." All these losses are a big loss to biodiversity, leading to an unbalanced structure of an ecosystem. Furthermore, coral reefs are known to be the most biodiversity—rich ecosystems (in species per unit area) in the world, more diverse even than tropical forests. Their health and resilience are in decline because of overfishing, pollution, disease, and climate change. Caribbean coral reefs have been reduced by 80% in three decades. As a direct result, revenues from dive tourism (close to 20% of total tourism revenue) have declined and are predicted to lose up to US$ 300 million per year. That is more than twice as much as losses in the heavily impacted fisheries sector (UNEP February 2008).

Figure 4.1: On the left unspoiled Coral reefs at Los Roques Islands coasts of Venezuela, on the right bleached coral reefs (corals expelled algae given stress) in Barbados.

(Source: tourist photographs from diving tourists provided by Eviatour C. A.)

The underlying explanation for this situation is that in 1983, following several centuries of overfishing of herbivores, there was a sudden switch from coral to the algal domination of Jamaican reef systems. This left the control of algal cover almost entirely to a single species of sea urchin, whose populations collapsed when exposed to a species-specific pathogen. When the sea urchin population collapsed, the reefs shifted (apparently irreversibly) to a new state with a little capacity to support fisheries. This is an excellent example of the insurance value in biologically diverse ecosystems. The reduction in herbivore diversity had no immediate effect until the sea urchin population plummeted, illustrating how vulnerable the system had become due to its dependence on a single species

4.3.5 Global Climate Change

Both climate variability and climate change cause biodiversity loss. Species and populations may be lost permanently if they are not provided with enough time to adapt to changing climatic conditions. The main reason behind biodiversity loss is the changes done to the ecosystem due to extensive land-use. Activities on land like agriculture, urban spread, and transportation infrastructure are some of the basic reasons for biodiversity loss. The urban spread is affecting the land through the transformation of urban-rural linkages. According to Gartner, "Urban life-styles tend to raise consumption expectations and that 60% of the world's population will be urban by 2025, the rural–urban linkage or the urban "ecological footprint" is critical to land change assessments." According to many climatologists, "the greenhouse effect is likely to raise world temperatures by about 2°C by 2030, meaning that sea levels will rise by around 30–50 cm by this time. Global warming, coupled with human population growth and accelerating rates of resource use will bring further losses in biological diversity." Due to this, vast areas of the globe would get flooded leading to biodiversity loss, unbalanced ecological systems, and also the loss of human life. An example for climatic change is, heating of the earth's surface affects biodiversity because it endangers all the species that adapted to the cold due to the latitude (the Polar species) or the altitude (mountain species).

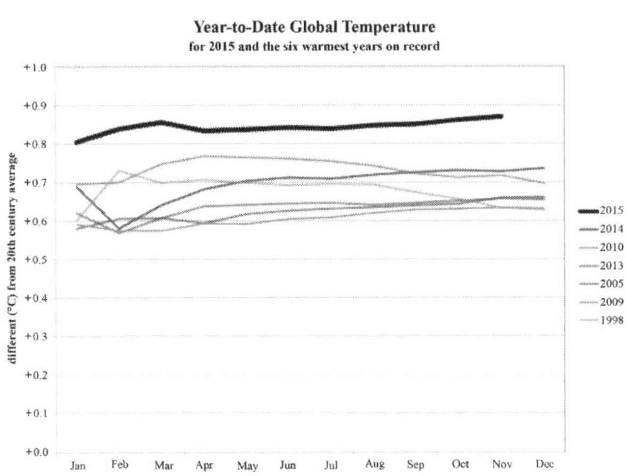

Figure 4.2: Rise in global temperature related to climate change

(From: https://collapseofindustrialcivilization.com/tag/global-famine/)

4.3.6 Agricultural Expansion

Apart from the urban expansion, agricultural expansion is another most important proximate cause of land-use that is changing biodiversity at a global level, followed by infrastructure development and deforestation. Population growth and consequently increased demand for food and fuel production are some of the primary drivers to convert nature into land for agricultural use. Habitat destruction is another main reason behind biodiversity loss.

4.3.7 Natural calamities:

Natural calamities, such as floods, draught, forest fires, earth-quakes, volcanic eruptions, epidemics, and so forth takes a heavy toll on the life of plants, animals, and humans. But it also affects the biodiversity in a very adverse manner. Floods have now become frequent in moist tropical regions of the world. They are causing a big deal of loss to the ground vegetation, loss of animals and erode away many soil nutrients from the ground. Also, there are regions where the failure of monsoon in succession for two or three years dries up ground vegetation, creating a situation of drought, due to which trees and plants also get affected.

Figure 4.3: Forest fire in the savannas of Africa affecting the endogenous species of elephants.

Source: https://static.independent.co.uk/s3fs-public/styles/article_small/public/thumbnails/image/2017/11/08/12/sanctuary-wildlife-photography-awards.jpg

With plant life animals also suffer. Another big natural calamity is the forest fires in densely wooded localities that often reducesa large number of plant and animal species to ashes and so do earthquakes. The below figure shows the wildfire erupted in Russia in the year 2010. Almost 1,800 animals and more than 100 people died in this incidence. Almost 2,000 houses got destroyed due to this calamity. According to the report flashed by the Fire department of that particular region, "Temperatures up to 100°F have exacerbated forest and peat bog fires across Russia's central and western regions."

Volcanic eruptions are another natural calamity as they destroy plant and animal life including the areas surrounded near that particular region. Epidemics sometimes destroy large portions of a natural population. In nature, such episodes are usually confined to specific plant or animal populations as the pathogen is often specific to particular species or group of species.

To summarize, it is identified that human-induced disturbances to the ecosystem, whether direct or underlying effects, are the major threat to biodiversity.

The indirect or underlying causes of biodiversity loss can be summarized in Table 4.1:

Table 4.1: Indirect or underlying Causes of Biodiversity Loss

Source: http://ec.europa.eu/environment/enveco/biodiversity/pdf/causes_biodiv_loss.pdf

Underlying Causes		
Demographic Factors	Economic and Market Factors	Institutional Factors
Population Trends that includes growth and density of population	Economic size and growth that include inequality, economic structure, economic footprint, GDP growth that results in loss of biodiversity	Formal Policies that includes policies on credits and economic development

Behavior of people towards biodiversity that includes attitude towards biodiversity, rent seeking and mitigation	Microeconomic Factors that include globalization and economic development models	Policy Climate that includes mismanagement and corruption
Public values, attitude, and beliefs that includes frontier mentality and indifferences about biodiversity	Trade-related factors that include IPR that is associated with trade agreements	Property rights that include titling and landraces

There are different direct and underlying causes that play a noteworthy role in biodiversity loss depending on the ecosystem under review. Change in habitat, overexploitation, pollution caused due to nitrogen and phosphorus have a negative influence on biodiversity loss in the forest ecosystem. In addition to this, both changes in habitat and pollution are also drivers for the loss of biodiversity for coastal ecosystems and inland wetlands. Moreover, invasive species are among other drivers that influence the biodiversity loss to a great extent in biomes. On the other hand, overexploitation is the most dominant driver of loss of biodiversity, which is followed by a change in habitat.

When there is lack of appropriate structure of governance, the marine areas suffer the most and become vulnerable to negative influence from economic growth. The direct effects of climatic change and water pollution threaten the ecosystem of freshwater. The direct cause of biodiversity loss that causes a threat to the coastal ecosystem is pollution and eutrophication. On the other hand, the underlying cause of threat to the coastal ecosystem is growing in population, transport, overpopulation, growth in the economy and international trade. On the contrary, the direct cause that threatens forest ecosystem is a change in climate, infrastructure, wood fuel, and extraction of timber and land conversion. The underlying cause of forest ecosystem includes increase in food demand, growth in the economy, and development in the international biofuel market.

4.4 IMPACT OR EFFECTS ON HUMAN BEINGS

It is evident that continuous biodiversity loss will have a substantial effect and severe consequences on global health and wellbeing of humans as well as global economies. The poor population across the world will be more likely to bear with the direct and instantaneous cost of loss of biodiversity.

Biodiversity is the major factor that helps in maintaining the considerate services of the ecosystem on which the development of economies is dependent as biodiversity helps in conservation of soil and purification of water. Biodiversity also controls diseases and reduces the vulnerability to various types of natural disasters such as landslides, droughts, and floods.

In addition to this, loss of biodiversity also worsens the major problem of poverty, which is itself a threat to biodiversity. Ecosystem services provided to human are the prime elements for paying attention so that a balance could be created in the ecosystem and global biodiversity can be maintained. To conserve biodiversity and maintain a balance between ecosystems, it is important to understand the link between biodiversity, ecosystem services, and human wellbeing.

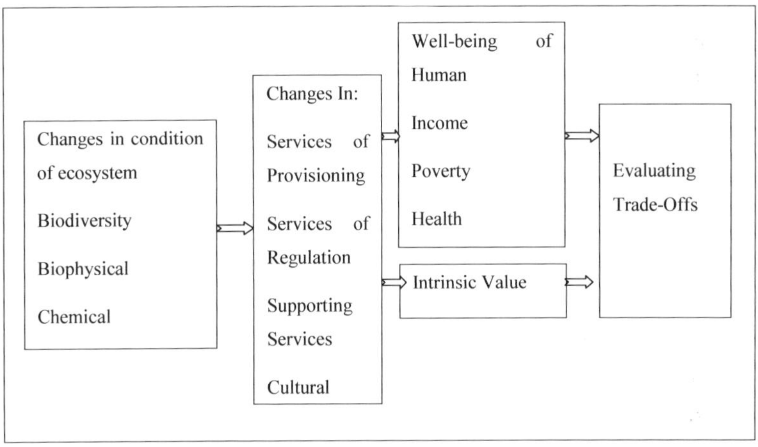

Figure 4.4: Link Between Biodiversity, Ecosystem Services, and Human Well-being.

Source: modified from: http://ec.europa.eu/environment/enveco/biodiversity/pdf/causes_biodiv_loss.pdf

There has been a rapid growth in the world population in the last century. According to a report by US Census Bureau, the total population of the world has increased and almost doubled in 30 years from the year 1959 to 1999 from 3 billion to 6 billion, respectively. It is also projected that the global population will continue to double up to 2030 and continue to rise throughout the 21st century. This growth in global population will ultimately increase the demand for forest and agricultural products. Although the growth of population is already a driver of loss of biodiversity further increase in population will increase the pressure on biodiversity. A research study reveals that only population growth alone cannot provide the explanation of biodiversity impact, but economic size and structure should be given equal importance. The continuous increase in the world population has the direct influence on the increase in the demand of the existing natural resources that can lead to the reaping of natural resource at unmanageable rates. It is quite evident that the population in urban areas is directly dependent on the people living in rural areas for food and fiber manufacture that is a source of survival.

Moreover, people in urban areas consume food and use fibers at a higher intensity in comparison to rural population. Due to increased population pressure, rural people will more likely move towards less occupied or less dense regions that will affect their livelihood opportunities. Moving towards less dense areas may also lead to environmental deterioration in these regions that will further impact the suitability of habitat and negatively impact the biodiversity in that area.

In addition to this, ecosystem imbalance and loss of biodiversity occur due to failure in controlling the adequate use of water, resources, and land. Furthermore, in developing nations, the environmental programs are not paid prioritized attention due to which such programs suffer from budget restraints that lead to environmental consequences. When there is a budget constraint in social services, the poor population is more likely to be affected because they heavily rely on the natural products and resources such as food, and wood fuel.

It is considered however that governmental and institutional factors are the key drivers that can help in improving the policies for biodiversity. Though widespread governmental framework is present in many nations, the biodiversity policy is not successful enough. There are several factors that result in the failure of the biodiversity policy such as international cooperation, stakeholder involvement, lack of government's political will and implementation of the effective measures. These factors that result in policy failure usually arise in developing nations. The increasing economic activities in both the developed and developing nations also imposes a negative impact on biodiversity as it affects the habitat destruction and increases soil, air, and water pollution.

In the present environment, the global consumption and production patterns are supported by the ecosystem surrounding the world. The resilience of natural and artificial ecosystems is affected by several types of policy ranging from transport to agriculture, energy to cultural well-being. There are many effects on the ecosystem from food, water, sea, and health, which can be illustrated with the help of various examples.

4.4.1 Increasing prices of Food and Land based on Biodiversity Loss

Prices of food are rising continuously due to which protests have been provoked in several nations. In the year 2007, thousands of people marched through the Mexican City streets in the month of February to determine that the increasing demand for biofuel in the US has increased the cost of corn by 400%. This has affected the production of corn, which is used to make tortillas.

Biofuels generate much debate

Bioenergy can play an important role in combating climate change, specifically if biomass is used for heat and electricity generation. However, biofuels also are another source of competition for scarce land, and the scale of potential land conversion for agro-fuels is extraordinary. The International Monetary Fund reports that "although biofuels still account for only 1. 5% of the global liquid fuels supply, they accounted for almost half of the increase in consumption of major food crops in 2006-2007, mostly because of corn-based ethanol produced in the US". Reports indicate that this pattern could be replicated elsewhere in the world.

Figure 4.5: Biofuel Create Dispute.

Source: http://ec.europa.eu/environment/nature/biodiversity/economics/pdf/teeb_report.pdf

There are several causes that promote an increase in the prices of food such as rising demand for food, which requires more land to cultivate food, increase in the price of energy and demand for biofuels. According to FAO in the year 2007 calculation of food price index, which shows an increase in food prices by 40% in 2007, was recorded only 9% in 2009. There was a drastic rise in food prices at the beginning of 2008 again. It is determined that when the basic commodity demand increases, it imposes a rising pressure to alter ecosystem to farmland along with mounting the production intensity from converted land. The world has already shifted to high consumption of meat that is an important cause of deforestation across the world.

It is estimated that billions of people rely on fisheries as their main source of food and protein, but half of the fisheries are subjugated, and few are overexploited. The fisheries are not only used as a source of food or protein, but it is also used to feed poultry and to make fish oil. Hunting marine wild fisheries influence the biodiversity of the ocean. If the biodiversity loss continues to rise, there will be a disastrous effect on the supply of seafood to the humans and the economy because fishery diversity is disastrously declining.

The fundamental apprehension of mankind is the conservation of biodiversity that helps in facilitating a sustainable attitude. The current state of biodiversity loss is at increasing pace, and it will continue to grow in coming decades. The key drivers of biodiversity loss are increasing global population and economic activities. The requirement of energy, wood, fodder, water, and food will lead to the diminished and unmanageable use of natural resources. Moreover, the further negative impacts will be imposed by the change in climate, deposition of nitrogen, fragmentation, and increasing infrastructure. As a result of all these factors biodiversity is estimated to decrease from 70% in the year 2020 to 63% by 2050.

4.4.2 Decreasing Water Supply as a Result of Biodiversity Loss

There is growing water stress worldwide in terms of water supply and quality of water. Due to climate change, the water supplies are at risk, which influence the irrigation of crops especially in the countries that largely depends on agriculture. To maintain the hydrological cycle, forests and wetlands play a significant role in terms of absorption and retention of the water. However, this is possible only when the ecosystem is safeguarded as it plays an important role in determining whether there is sufficient water to drink.

Forests and peatlands are very important in maintaining water services in rivers basins. The higher structure of the biomass the higher interception of the rainfall occurs and the replenishment of aquifers. Eliminating forests cover leads to increasing surficial runoff and water loss also by increasing evaporation from the river basin.

4.4.3 Effect on Health as a Result of Biodiversity Loss

Traditionally, for thousands of years, our ancients, as well as present population, know the medicinal value of certain plants. Moreover, biodiversity has helped humans to understand their body structure intensely. Thus, the ecosystem can be said to provide immense health benefits, but increasing threat to biodiversity is a threat to human beings themselves. This is because there is a direct association between healthcare and biodiversity.

- It is estimated that almost half of the synthetic drugs are prepared organically from extracts of certain plants or animals. For example, every 10 of the 25 maximum selling drugs in the United States have a natural origin.
- 42% of the entire range of anti-cancer drugs that are available in the market are prepared naturally, and 34% are prepared semi-naturally.
- Out of the total of 30,000 higher species of plants in China, 5,000 plant species are used for healing purposes.
- It is estimated that approximately three-quarters of the entire population in the world depends on the natural and traditional remedies.
- In addition to this, it is determined that the total drug turnover derived from the genetic resources is estimated to be between the US $75 billion to the US $150 billion in the year 1997 in the United States.
- With the discovery of gingko tree, the cardiovascular diseases are effectively treated, which accounts for a total turnover of US $ 360 million each year.

Table 4.2: Commonly used drugs and their plant sources

Source: https://www.ugc.ac.in/oldpdf/modelcurriculum/Chapter4.pdf

S. No.	Drug	Plant Source	Use
1	Atropine	Belladonna	Anticholinergic: reduces intestinal pain in diarrhea
2	Bromelain	Pineapple	Controls tissue inflammation due to infection
3	Caffeine	Tea, Coffee	Stimulant of the central nervous system
4	Camphor	Camphor tree	Rubefacient: increases local blood supply
5	Cocaine	Coca shrub *Erithroxylum coca*	Analgesic and local anesthetic: reduces pain and prevents pain during surgery
6	Codeine	Opium poppy	Analgesic: reduces pain
7	Morphine	Opium poppy	Analgesic: controls pain
8	Colchicine	Autumn crocus	Anticancer agent
9	Digitoxin	Common fox-glove	Cardiac stimulant used in heart diseases
10	Diosgenin	Wild yams	Source of female contraceptive: prevents pregnancy
11	L-Dopa	Velvet bean	Controls Parkinson's disease that leads to jerky movements of the hands
12	Ergotamine	Smut-of-rye or ergot	Control of hemorrhage and migraine headaches
13	Glaziovine	*Ocotea glaziovii*	Antidepressant: Elevates mood of depressed patients
14	Gossypol	Cotton	Male contraceptive
15	Indicine N-oxide	*Heliotropium indicum*	Anticancer agent
16	Menthol	Mint	Rubefacient: increases local blood supply and reduces pain on local application
17	Monocrotaline	*Cotolaria sessiliflora*	Anticancer agent
18	Papain	Papaya (*Carica papaya*)	Dissolves excess protein and mucus, during digestion
19	Penicillin	*Penicillium fungi*	General antibiotic kills bacteria and controls infection by various micro-organisms present in these antibiotics as antibacterial organisms.

20	Quinine	*Cinchona* spp.	Antimalarial
21	Reserpine	Indian snake-root	Reduces high blood pressure
22	Scopol-amine	Thorn apple	Sedative
23	Taxol	Pacific yew	Anticancer (ovarian)
24	Vinblas-tine, vincristine	Rosy peri-winkle (*Vinca rosea*) (Sadaphali)	Anticancer agent: Controls cancer in children.

All the above mentions health benefits are gained from the growing knowledge about ecosystem and biodiversity. Despite several health benefits, plants are continuously vanishing at an alarming rate. Therefore, urgent actions need to be taken. It is revealed in a recent research study that several species of medicinal plants are at extinction. The plant species specifically that have medicinal properties are usually found in tropical regions, and these plants are the sources of forming many new drugs. People in developing nations are highly motivated to conserve natural habitat because they are the one who is more likely to get benefit from medicinal plants and their resulting drugs. This is because people in developing nations are able to afford these drugs and medicines and these drugs are easily made available in such nations.

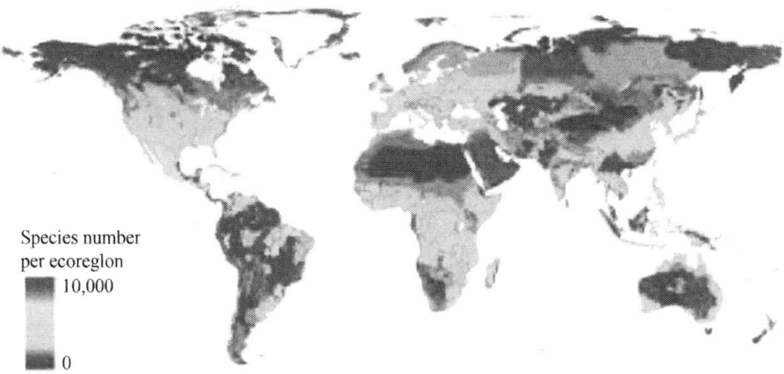

Figure 4.6: Species of Plants per Ecoregion.

Source: https://www.google.co.in/url? sa=t&rct=j&q=&esrc=s&source=web&cd=15&cad=rja&uact=8&ved=0ahUKEwjN_5L-gePWAhUDvI8KHfQwDtA4ChAWCDwwBA&url=http%3A%2F%2Fec.europa.eu%2Fenvironment%2Fn

ature%2Fbiodiversity%2Feconomics%2Fpdf%2Fteeb_report.pdf&usg=AOvVaw1srEc4p1In8aEZ7hvrfUsr

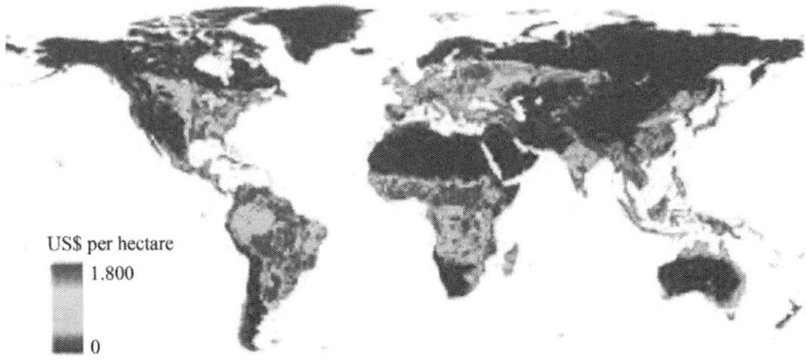

Figure 4.7: Agricultural Return per Ecoregion.

Source: https://www.google.co.in/url? sa=t&rct=j&q=&esrc=s&source=web&cd=15&cad=rja&uact=8&ved=0ahUKEwjN_5L-gePWAhUDvI8KHfQwDtA4ChAWCDwwBA&url=http%3A%2F%2Fec.europa.eu%2Fenvironment%2Fnature%2Fbiodiversity%2Feconomics%2Fpdf%2Fteeb_report.pdf&usg=AOvVaw1srEc4p1In8aEZ7hvrfUsr

4.4.4 Climate Change and Biodiversity

Change in climate and biodiversity are directly related to a little change in climate can make biodiversity vulnerable to climate. For instance, the cycle of El Niño-La Niña is one of the significant examples that depict the vulnerability of biodiversity to climate. El Niño-La Niña is a natural activity that is caused due to climatic change. This is a weather pattern that is caused when the temperature and atmosphere above the Pacific Ocean changes. Between 1976 and 1998, there was small temperature upsurge in the sea surface, which led to various phenomena across the world. The El Niño-La Niña cycle caused permanent damages worldwide that includes:

- Forests were burned due to El Nino-La Nina activity that has long-term effects as those forests are considered unrecoverable.
- During that phase, the temperature of ocean surface rose from an average of 19–25°C.
- It also caused modifications towards heat-tolerant species existing inside corals.
- The jet stream also shifted northward due to El Nino-La Nina effect.

Furthermore, biodiversity losses also contribute to change in climatic conditions in several multifaceted ways. It is revealed that change in land use pattern and over-harvesting have prompted economic and social changes that further leads to more dependence on carbon. Drainage in wetlands results in a reduction in evaporation and drier soil that further cause loss of carbon. However, it is estimated that climate change can fasten the release of carbon from the soil, which further contributes to higher emission and concentration of greenhouse gas in the atmosphere. This emission of a greater amount of carbon in the air causes air pollution, which affects the human health.

4.4.5 Impact on the Poor

There are striking aspects of consequences that occur as a result of the loss of biodiversity that has several unrecognized effects on the poor population. For example, change in climatic condition causes floods, droughts, and earthquakes that directly affected the poor population. It is evident that ecosystem services such as food, water, fisheries and other ecosystem benefits are not equally shared across the world. The poor women are considered most prone to deforestation and ecological imbalance. This is because there are several poor female labors who walk down to forests to collect forest products but due to depleting trees and medicinal plants they are unable in accessing forest produce. This inability reduces their income level, affects their physical health and also increases labor.

Due to depleting biodiversity, the poor population across the world is among the most vulnerable ones who will likely bear the immediate and direct effect of such loss. Biodiversity loss worsens poverty, and in a similar manner, poverty imposes a major threat to biodiversity. Deforestation leads to global warming, which reduces water regulation, which further affects the livelihood of the poor people across the world. It is already discussed that poor people are heavily dependent on biodiversity and ecosystem services thus, loss of biodiversity will negatively affect the livelihood of the poor people.

4.5 STABILITY OF ECOSYSTEM, THE NEED FOR CONSERVATION

The foundation of ecosystem services is the biodiversity and the welfare of human beings is associated with these services. The layer of a living organism is much more complex than the features of the earth because it is living organisms that experience change when the surface and sea on the earth influences. This is because living organisms occupy the surface and seas of the earth and are influenced the most when there is any change on earth. The biosphere is the layer where the living organism lives with the cooperative metabolic activities of the microbes, plants, and animals unite the hydrosphere, atmosphere, and geosphere in both physical and chemical manner. There is an escalating indication that biodiversity helps in increasing the stability of ecosystem processes that influence the environment.

Pollination is an ecosystem service that increases stability processes at the ecosystem level; it is provided by a wide variety of organisms, and is essential to the production of many of the world's food crops.

Though most of the biodiversity hotspots that are critically endangered are located in developing nations chiefly in the southern hemisphere. However, it is also evident that biodiversity loss is also noted in some parts of developed nations such as United States of America, Australia, and the European Union. This shows that biodiversity loss is increasing at an alarming rate and revolutionary actions should be taken urgently.

A common solution must be adopted to safeguard both ecosystem services and species. The two primary objectives of conserving nature are to conserve the biological diversity of the earth and to secure ecosystem services benefits. We have shown how the services offered by ecosystem provide a direct support to millions of people living in poverty therefore, it is crucial to protect ecosystem not only for poverty reduction but for economic development as well. However, biodiversity loss and a decline in ecosystem services might cause huge cost for society because it influences various regulatory and provisional services such as a decline in production of food, water regulation and resilience in climatic change. Loss of all these services will impact the human beings on a large scale in the near future, therefore, the rate of loss requires urgent actions to be taken for biodiversity conservation. Finally, conserving biodiversity is also expensive as it might require a restriction on different opportunities that might affect economic development.

Although, a huge amount of money is invested globally for biodiversity conservation ranging between the US $8 billion and the US $10 billion annually. It is estimated that in the next 30 years the US $28 billion might be required annually to conserve biodiversity. This cost estimation is done by including management and acquisition costs of the present and future biodiversity reserve spots. It is projected that conservation costs vary between different locations because their cost and economic structure area are different. It is considered that establishing and managing protected areas in developed nations are expensive in comparison to the developing nations. Moreover, the cost of accomplishing a given target of conserving biodiversity depends highly on selected policies and their designs.

It is noteworthy that different species all around the world are at the extinction stage at the fastest rate, which is mostly due to human activities. Few conservation organizations projected that species are moving towards extinction at the frequency of one in every 20 minutes. There is rapid species loss in the present state that is estimated to be ranging between 1,000 and 10,000 times faster than the natural rate of species extinction. Furthermore, it is also estimated that if human activities and business continue as usual, 0.01–0.1% of species will head towards extinction every year.

However, a Red list of threatened species of 2009 according to The International Union for Conservation of Nature (IUCN) was generated and states that 17,291 species out of the total 47,677 species were deemed to be at the risk of extinction. Thus, biodiversity conservation is essential as it provides extensive advantages that help in meeting the immediate human requirements such as clean drinking water, natural food, and will also protect from unstable climate, flood, droughts, and storms. The biodiversity loss is considered as hazardous and its penalties are immediate such as:

- Loss of biodiversity results in lack of opportunities that affect livelihood, health, and education
- The human activities like fish hunting are decreasing the number of fisheries, which results in biodiversity loss. Humans are largely dependent on seafood and this loss further results in fewer fishes in the sea, which means that there will be less seafood for human beings.
- Another immediate consequence of biodiversity loss is lack of clean drinking water
- Biodiversity loss also results in the lack of forest resources and ecosystem services such as food, wood fuel and plants for medicine.
- In the long-run, biodiversity loss also means that there will be less income and wages for poor community people.

Thus, immediate solutions need to be implemented for conserving biodiversity. there are several initiatives that are implemented for enhancement of sustainable development and for promoting biodiversity conservation all around the world. These initiatives include:

- The first and foremost initiative is the "Green Economy Initiative" that has positively influenced the biodiversity conservation. This initiative has enhanced the conversion to a low-carbon economy along with resource-efficient economy through which the economy is able to meet several challenges and deliver numerous opportunities.
- Another initiative is the protection of biodiversity areas or protected areas. Through this initiative, promotion of increasing the connectivity, number, and size of biodiverse areas is performed. In addition to this, protected areas of both land and sea are focused.
- Moreover, an increased attention is paid on the application of Multilateral Environmental Agreements in terms of biodiversity conservation that includes:
    1. The Convention on Migratory Species (CMS)
    2. The Convention on Biological Diversity (CBD)
    3. The Convention on International Trade in Endangered (CITES)
    4. The Ramsar Convention on Wetlands and The World Heritage Convention
- Increasing the program initiatives that help in relating people and their livelihood to biodiversity. This will raise the awareness among the people about the biodiversity and they will take some actions to conserve it. CEPA (Communication, Education, and Public Awareness) programs is an example through which this aim can be achieved.
- In addition to this, public authorities should identify and create opportunities for rural enterprises that are based on biodiversity such as eco-tourism and bio-prospecting which will increase local employment and thus the environment, species and their habitat will be contributed.

REFERENCES

1. Benn, J. (2010). *What is biodiversity come with us on a journey*. United Nations Environment Programme.
2. Bharucha, E. (2004). Textbook for Environmental Studies for Undergraduate Courses of all Branches of Higher Education. New Delhi: University Grants Commission.
3. Bruford, M. (2017). *Decline in genetic diversity of black rhino*. Cardiff University.
4. Cardinale, B. J., Duffy, J. E., Gonzalez, A., Hooper, D. U., Perrings, C., Venail P., Narwani, A., MacE, G. M., Tilman, D., Wardle, D. A., Kinzig, A. P., Daily, G. C., Loreau, M., Grace, J. B., Larigauderie, A., Srivastava, D. S., & Naeem, N. (2012). *Biodiversity loss and its impact on humanity*. *Nature* (Macmillan Publishers Limited), *486*, 59-67.
5. Cheetah Conservation Fund. (2016). *About the cheetah: Genetic diversity*. Retrieved October 12, 2017 from https://cheetah.org/about-the-cheetah/genetic-diversity/
6. Secretariat of the Convention on Biological Diversity. (2001). *Handbook of the convention on biological diversity*. Earthscan/James & James. Costanza, R. (1992). *Ecological economics: The science and management of sustainability.* Columbia University Press.
7. Dimas, S., & Gabriel, S. (2008). *The economics of ecosystem and biodiversity*. Cambridge, UK: European Commission.
8. IUCN – International Union for the Conservation of Nature. (2008). *Community conserved areas: A bold new frontier for conservation.* Retrieved October 10, 2017, from www.iucn.org/themes/ceesp/CCA/Index.html
9. Johnston, I. (2016). *Rate of species decline 'no longer within safe limit' for humans, experts warn: 'Until and unless we can bring biodiversity back up, we're playing ecological roulette'*. Independent.
10. Maheshwari, V. K. (2016). *Biodiversity: The variability among living organisms*. Retrieved October 12, 2017 from http://www.vkmaheshwari.com/WP/? p=2254
11. Meyer, A. (2016). *Smithsonian study reveals decline of genetic diversity in wild cheetahs*. Retrieved October 12, 2017 from http://insider.si.edu/2016/06/smithsonian-study-reveals-precipitous-decline-genetic-diversity-wild-cheetahs/
12. Mittermeier, R. A., Turner, W. R., Larsen, F. W., Brooks, T. M., &

Gascon, C. (2011). Global biodiversity conservation: The critical role of hotspots. In F. E. Zachos, J. C. Habel (Eds.), *Biodiversity hotspots: Distribution and protection of conservation priority areas* (pp. 3-22). Heidelberg: Springer.

13. Myers, N. (1988). Threatened biotas: "hotspots" in tropical forests. *Environmentalist*, *8*(3), 187-208.
14. Rubidge, E. M., Patton, J. L., Lim, M., Burton, A. C., Brashares, J. S., & Moritz, C. (2012). Climate-induced range contraction drives genetic erosion in an alpine mammal. *Nature Climate Change*, *2*, 285-288.
15. Slingenberg, A., Braat, L., Windt, H. V. D., Eichler, L., & Turner, K. (2009). *Study on understanding the causes of biodiversity loss and the policy assessment framework*. Brussels: European Commission Directorate-General for Environment.
16. Sofo, A. (2011). *Biodiversity*. Rijeka, Croatia: InTech.
17. The Organisation for Economic Co-operation and Development (OECD). (2014). *Biodiversity and ecosystems*. Retrieved October 10, 2017, from http://www.oecd.org/env/resources/OECD-work-on-biodiversity-and-ecosystems.pdf
18. Turnet, W. R., Brandon, K., Brooks, T. M., Costanza, R., Fonseca, G. A. B., & Portela, R. (2007). Global conservation of biodiversity and ecosystem services. *Bioscience*, *57*(10), 868-873.
19. Whigham, N. (2016). *The mass extinction event going unnoticed as the planet's biodiversity dwindles*. News.com.au.
20. World Conservation Monitoring Centre. (1992). Global diversity: Status of the Earth's living resources. Chapman & Hall.
21. Yang, S. (2012). *Yosemite's alpine chipmunks take genetic hit from climate change. Berkeley News*. Retrieved from http://newscenter.berkeley.edu/2012/02/19/climate-change-erodes-alpine-chipmunkgenetic-diversity/

5
CHAPTER

ECONOMIC SIGNIFICANCE OF BIODIVERSITY

"Biodiversity starts in the distant past and it points toward the future."

—Frans Lanting

CONTENTS

ABSTRACT

Biodiversity is one of the most important sources of economic wealth for almost all the regions of the world. The economic significance of biodiversity can be measured by understanding the total economic value of all the services rendered by the ecosystem to the living organisms. There are many sectors that are directly or indirectly causing a huge loss to biodiversity due to its over-use of ecological services. In this chapter, all the sectors, which contribute towards the economy of a region through ecological services, have been discussed in detail. Apart from that the economic valuation of biodiversity is also discussed along with the economic drivers causing a loss to biodiversity.

5.1 INTRODUCTION

According to many economists, it is believed that biodiversity is very much valuable to the society. If there is greater biodiversity, then living organisms would be in a better situation. Any choice made by the living organism directly affects the biodiversity in some way or the other. For example, any wrong decision made, which has a negative impact on biodiversity and further leads to its loss then this would also adversely affect the ecosystem and all the living organisms. Cutting down of trees for agriculture, or for construction purpose, draining wetlands for the construction of infrastructure causes a huge depletion in biodiversity. Economist claims that biodiversity is a valuable and scarce resource. The analysis is done in the ways societies make choices regarding the resources and services provided by the ecosystem. Below in this chapter, a brief discussion has been done to learn about various economic aspects of biodiversity. The value of biodiversity in context to economics, and how it could be measured are discussed in detail.

The economic value of biodiversity helps the humans to understand that what all policies should be formulated for the sustainable use of resources available and both the strengths and weaknesses of the economics approach are required for reducing biodiversity loss. Biodiversity or biological diversity embrace the whole of "Life on Earth." It describes the variety and also the variability of all the living organisms in a given framework. If there is any sort of decline in biodiversity, then the changes occurred are directly or indirectly related to the reduction and simplification of biological heterogeneity, from individuals to various regions. Basically, biological resources are those important resources of biodiversity, which helps in maintaining and organizing the necessities of human beings in a systematic manner. This is considered as the anthropocentric view of biological resources, which offers a flexible and transparent way to analyze alternative value paradigms in context to economic factors. For many environmental economists' intrinsic values are more appropriate for taking decisions related to the conservation of resources, but these intrinsic value of biodiversity is difficult to measure.

All the services rendered by the ecosystem like, agriculture, forestry and fisheries products, stable natural hydrological cycles, fertile soils, a balanced climate are dependent on practices taken up for the conservation of biological diversity. Production of food is dependent on biodiversity for a variety of food plants, pollination, nutrient provisions, disease prevention and control, and so forth. Both medicinal plants and manufactured pharmaceuticals rely on biodiversity. When there is a decline in biodiversity, it leads to an increment in the transmission of diseases to humans, thus increasing the healthcare costs of an individual. Also, the outdoor tourism industry relies on biodiversity to create and maintain the records of all visitors and what is the contribution of tourism industry in growth of fishing and hunting industry.

5.2 SECTORS CONTRIBUTING TOWARDS THE ECONOMY OF A COUNTRY

There are many sectors that contribute towards the economy of a nation by using natural resources. The sectors discussed below are highly dependent on renewable resources available in the ecosystem. These resources used by these sectors are the components of biodiversity and are also the services rendered by the ecosystem.

5.2.1 Agriculture

Agriculture is one of the major sectors that contribute towards the Gross Domestic Product of any country at a maximum level. It provides the export revenues, employment, and rural livelihoods to a particular region at a greater scale. According to a report flashed by UNDP (United Nations Development Program), the contribution of agricultural sector to United Nations' was really high. The report stated that, "its contribution to GDP over the period 2000–2007 averaged 9.6% for the region, while agricultural exports represented 44% of the region's total export value in 2007." Apart from United States, the contribution of agriculture exports in many countries like, Panama, Paraguay, and Nicaragua were more than 80% of the total commodity exported in 2007. The major crops that are generally exported globally are soybean, coffee, bananas, beat sugar, and so forth. In India almost 25% of population is involved in agricultural sector, whereas is Latin America, around 9% population is employed in the agriculture sector and agriculture is a primary source of income for rural households.

Figure 5.1: Agricultural practices.

Source: https://pixabay.com/en/farmer-tractor-agriculture-farm-880567/

5.2.2 Fisheries

Figure 5.2: Fisheries.

Source: http://metrovaartha.com/en/2017/07/06/action-plan-drawn-up-for-sustainable-development-of-marine-fisheries/

The fisheries sector is economically important in many countries, which contributes highly to GDP (Gross Domestic Product), food security, domestic income, foreign exchange revenue, and so forth to a region. The fiscal revenue of many countries is also somewhere dependent on fishery, which is an important component of aquatic ecosystem. According to UNDP report, "fisheries provided a revenue of around $2 billion to countries like, Chile, Mexico, Colombia, and Brazil, whereas, Venezuela, Panama, Argentina, Guyana, and Peru more than $100 million in 2004. In relative terms, fisheries are important at the national level in a wide range of countries in LAC, contributing more than 1% of GDP in at least 10 countries." Fisheries serve poor people residing in the coastal regions by providing them their livelihood.

5.2.3 Forestry

Forestry is one of the most important manufacturing sector for any country. There are many forest products that are regarded as the daily essential for the sustenance of human lives, for example, lumber and other solid wood products, and pulp and paper. According UNDP, "On average, logging and related processing activities in LAC contribute 2% to GDP. From 1990 to 2006, the forestry part of GDP in LAC rose from $30 billion to $40 billion, most of it from round wood production. In terms of job generation, FAO (2008) put the number of people employed in the round wood, and pulp and paper and wood processing activities at 1.5 million in 2006, or 0.75% of total employment in LAC."

Apart from the United States, the forest industry is one of Canada's most important manufacturing sectors. In 2016, it:

- accounted for about 7% of Canada's total exports,
- contributed almost $23 billion to Canada's economy,
- directly supported about 211,075 jobs across the country (rise in employment), including an estimated 9,700 Indigenous people,
- generated more than $1 billion in revenue (2015) for provincial and territorial governments.

Besides their direct contribution to GDP, forests have substantial potential to generate ongoing economic production based on ecological services from water to food, fiber, carbon sequestration, non-timber forest products (NTFP), and tourist attractions.

In Guyana there is a Sustainable Forestry Program with the participation of indigenous communities.

5.2.4 Tourism

Many species (plant and animal) attract a large number of tourist towards the region they are located in. tourism in context to biodiversity contributes a lot towards the economy of a particular region. The contribution of the tourism sector to GDP for United States as a whole range from about 2% in the larger countries of South America to almost 20% in the Caribbean, with Central America ranging in between. "Growth rates in LAC averaging 8% have consistently outstripped North American and global growth rates for 15 years. The Caribbean, with a reputation for conventional business as usual tourism, has seen growth rates dip in recent years." The tourism sector is considered as a major employer sector in many countries like, United States and India. Almost 5–9% of employees are employed in tourism sector.

5.2.5 Protected areas

Terrestrial and marine reserves provide crucial ecosystem services to each of the sectors discussed above. All the terrestrial and marine reserves comprise provisions related to water purification, clean water for irrigation, hydropower and consumption of commodities. Growing green markets are opening significant opportunities to protected area (PA)-related business. Through nature-based tourism, protected areas have brought the opportunities of employment, local development, and prosperity to small regions and contributes highly to the GDP of that particular region, and earn high amount of foreign exchanges. Under Sustainable ecosystem management, protected areas can help in eradicating poverty and also helps in bringing equality in various regions where people are now able to afford opportunities to build self-sufficiency.

5.3 ECONOMIC VALUE OF BIODIVERSITY

For efficient use of scarce resources available in the ecosystem, the economic analysis plays an important role in maintain the ecological balance. An important consequence of the environmental crisis is that biodiversity is increasingly recognized as an essential yet diminishing resource. When the supply of ecological resources was in stable situation, the use of economic resources was not taken into consideration. According to Freedman in 1995, "As long as the supply of these services was more than ample, there was no need to consider them in economic terms, and for this reason economic activity often caused them to be significantly degraded. An important step in sustaining these conventionally non-valuated resources is, therefore, to define them as goods and services, which can be quantified in economic terms."

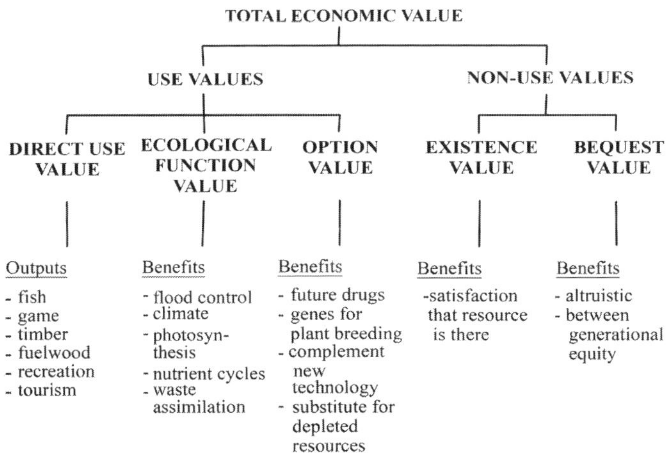

Figure 5.3: Nature of total economic value.

Source: http://www.csun.edu/~sd1229/444/2009/Reading/EdwardsaandAbivardi.pdf

In recent years, many economists have made an important contribution in defining that how important biodiversity and its resources are for the sustaining a healthy life. They have also helped in developing methods for assessing their value. Many economists like Groombridge, 1992; Filion and Adamowicz, 1994; Pearce and Moran, 1994; Young, 1995; "There is now a well-established classification of major categories of value from biodiversity. The most obvious value is from direct extractive uses, such as the production of timber or the collection of plants and animals for food. There may also be direct non-extractive uses of biodiversity, such as their importance for recreation or tourism." Tourism is a major category of international trade in services. In addition to receipts earned in destinations, international tourism also generated US$ 211 billion in exports through international passenger transport services rendered to non-residents in 2015, bringing the total value of tourism exports up to US$ 1.5 trillion, or US$ 4 billion a day on average.

Biodiversity contributes a lot to the Gross domestic product of almost every country and helps them to increase the national economy all around the world. According to the Secretariat of Convention on Biological

Diversity (SCBD), "the estimated economic potential of biodiversity in the pharmaceutical sector is about $640 billion of which 25–50% is derived directly from biological resources, globally. Herbal supplements, personal care products, and food products collectively have a market size of $65 billion, representing the largest "nature"-based market. The economic potential of coral reefs, on an average, at any given site include the following — an estimated $189,000/hectare/year in reducing natural disasters, tourism up to $ 1 million/hectare/year, genetic material and prospecting up to $57,000/hectare/year, fisheries up to $3,818/hectare/year."

The coral reefs have an economic potential of almost $1,250 million/hectare/year. These are the figures flashed by the researchers and economists collectively. For example, only 10% of potential coral reefs are required for the local people of any one particular region. Which makes around $125,000 per year/hectare out of the total value of coral reefs calculated. According to Indian currency, it is calculated as Rs. 61,25,000/hectare/year. If India has on an average 2,330 hectares of coral reefs, which comprises Rs. 14,27,12,50,000 of revenue per year from this simple but critical ecosystem as of now. The contributory figures can go up many-fold if we invest in the proper maintenance of this ecosystem.

Another example of the economic value of biodiversity is the conversion of land for further infrastructural development. In Southern Thailand, it was seen that mangroves were converted into shrimp farming, which gave them profitability of around $1220/hectare/year along with the subsidies provided for. Out of these $1220/hectare/year, $584 were generated from commercial profits from mangrove forests and the rest from raising fish nurseries. If the mangrove area is protected without conversion, the average yields from these measures are estimated as $10,821/hectare/year that includes the savings from storm protection in the area.

On the other hand, the interim report of the study titled The Economics of Ecosystems and Biodiversity (TEEB), mentions that "there are two challenges today with regard to ecosystems and biodiversity. Firstly, they are still learning the "nature of value," as they broaden our concept of "capital" to encompass human capital, social and natural capital. Secondly, they are still struggling to find the "value to nature." The report also claims that any sort of lack of valuation done in the economic factors of biodiversity is the underlying cause for the observed degradation of ecosystems and loss of biodiversity.

5.3.1 Policies for the Economic Sustenance

After taking a deep dive into the significance of economy with respect to biodiversity, five key solutions for better stewardship of natural capital have been realized. Thcse 5 key solutions include:

- Rewarding benefits of the ecological services through payments and markets,
- Reforming environmentally harmful subsidies,
- Developing local entrepreneurship,
- Addressing losses to natural capital through regulation and
- Pricing and investing in ecological infrastructure.

India is a mega-biodiverse country bequeathed with great natural capital, where ecologists should consider translating the natural capital into actual value. A few possible approaches in this regard are: Payments and markets should be made to work for the conservation of resources (natural) so that they are widely available for the poor. The misunderstanding that payments for ecosystems are parallel to privatizing the natural resources should be dispersed. One of the challenges will be to mainstream the costs and benefits of ecosystem services into local and regional development plans. Efforts are required to develop and implement sectoral policies, which would consider local markets as key to scaling up regional markets.

Another method for implementing a monetary value upon a natural resource and ecosystem services is by having a questionnaire session with different groups of people at different point of time. This is also a part of conducting research by the economists and ecologists for having a clear picture of the demand and supply of resources and services available for commercial purposes. These questions thus, address people's willingness to pay for biological diversity, for example through increased taxation. Such "Contingent Valuation" methods can be used to estimate not only hypothetical use values (for example, how much will people pay to be able to swim in clean rather than polluted water), but also non-use values. These categories of economic value are important for conservation because they force us to consider all of the benefits, which we obtain from wildlife, and not simply those that are tradeable.

5.3.2 Economic Valuation of Biodiversity

In Table 5.1, a common classification for environmental asset valuation is described in detail. The total economic value of an ecological resource comprises the use value and non-use value of resources by the living organisms.

Table 5.1: Categories of economic values attributed to environmental assets

(Source: https://www.cbd.int/financial/values/g-economicvalue-iucn.pdf)

	Use values	**Total economic value**	**Non-use values**	
Direct use	Indirect use	Option values	Bequest values	Existence values
Outputs directly consumable	Functional benefits	Future direct and indirect values	Use and Non-use value of environ-mental legacy	Value from knowledge of continued existence
Food, Biomass, Recreation, Health	Flood control, Storm protec-tion, Nutrient cycles	Biodiversity, Conserved habitats	Habitats, preven-tion of irreversible change	Habitats, Spe-cies, Genetic, Ecosystem

Use value can be described as a value that is drawn from the actual use of a resource available to the living organism, for example, cutting down of forest for the use of timber, fishing, and so forth. Use values are further divided into two fractions.

- The first is known as direct use values. The actual uses such as fishing, timber extraction, and so forth, are the main processes involved in direct use values.
- The second is known as indirect use values. It refers to the benefits that have been derived from ecosystem functions. Some of those functions are functions of forest to protect the watersheds, and option value, which is a value approximating an individual's willingness to pay to safeguard an asset for the option of using it at a future date. This is more like an insurance value.

Non-use values (NUV) are the ones that have been divided into two different values. The first value is known as bequest value, and the other value is termed as "passive" use value. Bequest value measures the advantages that an individual gets from the knowledge that others might benefit from a resource in future. Whereas, passive use values are not related to the current use or option values, deriving simply from the existence of any particular asset. Suppose, in total, we have:

$$TEV = UV + NUV = (DUV + IUV + OV) + (XV + BV)$$

Where,

TEV = Total economic value

UV = Use Value

DUV = Direct use value

IUV = Indirect use value

OV = Option value

BV = Bequest value

XV = Passive use value

This is one of the sizeable technique for the valuation of biological resources. For example, Table 5.2 illustrates the environmental functions of forests. It depicts the where and how each category belongs to and also the extent of overlap in valuation.

Table 5.2: Environmental functions of forests

Source: https://www.cbd.int/financial/values/g-economicvalue-iucn.pdf

Sources of materials and services	**Sink for wastes**	**General and life support**
Timber	Absorption of waste	Genetic pool
Fuelwood	Recycling waste	Climate regulation
Other business products	Watershed protection	Carbon fixing
Non-wood products	Protection of soil and its quality and making it resistant to soil erosion	Habitat for people, flora, and fauna
Agricultural production		Aesthetic, cultural and spiritual source

From the above table and equation, it is understood that the function between the direct use of timber and an indirect protection of watershed is mutually exclusive in nature. The TEV (total economic value) technique helps in distinguishing between the use values and non-use value resources. It is only the bequest value and existence value that are difficult to evaluate.

5.3.3 TEV and its Economic Importance

The technique for economic valuation on ecological resources has been adopted by many economists in the form of TEV (total economic value). But, it is not necessary that the valuation of resources through this criterion would be true every time. There are many factors that criticize the TEV technique.

- This technique does not claim to use all the values (economic values).
- There are many ecologists who claim that total economic value does not explain the entire economic storyline of biodiversity in detail.

According to Turner in 1992, all these techniques and values are "primary values." He further states that "they are essentially the system characteristics upon which all ecological functions are contingent. There cannot be a watershed protection function but for the underlying value of the system as a whole. There is, in some sense, a `glue' that holds everything together, and that glue has economic value. If this is true, then there is a total value to an ecosystem or ecological process that exceeds the sum of the values of the individual functions."

According to this discussion, three reasons for the importance of biological diversity are extracted. Those three reasons are:

1. The first reason is based on the concept of economic value. If biodiversity is economically important, we would expect this to show up in expressed willingness to pay for its conservation. Shortly, we will show that this is indeed the case.
2. Economic value measurements will understate `true' economic value because of the probable failure to measure primary life support functions. This kind of economic value is difficult to observe because it is unlikely to be recognized until some disastrous event has happened: landslides consequent upon deforestation, loss of fishing grounds due to pollution, and so on.
3. Economic value does not capture nor is it designed to capture intrinsic value.

5.4 ECONOMIC DRIVERS

Markets are not the sole policy makers, and they are not the ones who assign a monetary value to biodiversity. Almost all the economists are very much familiar with this above-mentioned statement. It is because of the policy intervention that market prices hardly fluctuate and reflect the losses incurred by them to the society as a whole arising from biodiversity degradation. It is because of this market failure that leads to the misuse of biodiversity in an unsustainable manner by an individual or an industry. To integrate proper knowledge into national economic infrastructure, efforts should be made to enable a highly informative and effective program and policy for the sustainable development of the biodiversity and its resources. This could be done by recognizing the opportunity cost of biodiversity loss in a more advanced manner. Below various economic factors have been discussed, which affect biodiversity in different ways.

5.4.1 Market Failure

"Missing markets" is a part of market failures or market "imperfections" in the external benefits generated by biodiversity conservation and also refer to a lack of adequate information for price setting. Market failures can be of a local nature, and it is also measured on a global level. Market failure is referred as an inability of markets to capture the economic benefits for the country and leads to biodiversity loss. This can also be seen as the inability of markets to capture the costs of converting ecologically valuable land to other uses and losing biodiversity in the process. The latter refers to the fact that biodiversity conservation yields benefits to beneficiaries external to where conservation practices take place.

5.4.2 Value and Economic Structure

A major problem in connection with losses of biodiversity in different biomes is that much of the biodiversity value is not directly usable. If the use of resources is direct then it becomes easier to evaluate, supervise, and control the use of natural resources for commercial purposes as compared to the indirect use. Indirect values such as genetic information and aesthetic appreciation are not easy to capture and evaluate because their contribution in the market and to the trade is not stable and gets scattered over regions, time and space. Because these biodiversity values are not owned, bought, or sold, they are often at a competitive disadvantage in a market economy, leaving them undervalued and overused.

On the other hand, solutions for such imperfections could be made by creating nature reserves or national parks (in-situ conservation practices) so that biodiversity remains unaffected by market forces and their changing trends. However, this type of solution is not always optimal and is a rather ecologically static approach. Thus, another solution for this could be made by capturing the value of biodiversity through specifying appropriate property rights for all its uses, which is a problematic process in its own right.

A lot of research has been done on the connection between the environmental change and socioeconomic factors of biodiversity in a general manner. These studies have typically examined the relationship between environmental change and gross domestic product (GDP) or GDP per capita. Some of the more recent studies have found that "higher rates of economic growth are associated with greater biodiversity loss. It is doubtful such a simple relationship can explain biodiversity trends."

There are many economic structures that help in understanding the properties that drive biodiversity loss. The structure of the economic activities and the way a society is unequal is one factor that helps in determining the biodiversity loss due to economic structure. In a recent study, an analysis of data from 50 countries compared different socioeconomic models' ability to predict biodiversity loss. The results indicated that "economic inequality is an important predictor of biodiversity loss. Specifically, statistical comparisons revealed that the "economic footprint" (the size of the economy relative to the country area), together with "income inequality" were the best predictors of the proportion of threatened species." For instance, most of the countries in South East Asia are developing countries, and hold a large number of people living below national poverty lines. Marine resources still directly sustain the lives of over 120 million people, while overfishing is regarded the most pervasive threat to coral reef biodiversity.

5.4.3 Biodiversity Loss due to Trade

The commercial significance related to the genetic resources stored in various living as well as non-living organisms of the ecological system has increased with the passage of time. This is because of the intellectual property rights that have been assigned in context to the trade agreements made by the businessmen at an international level. According to critics of the expansion of international trade agreements, "this process creates incentives that may reduce biodiversity. Countries are deemed to be under pressure to change their IPR laws to conform to the TRIPS agreement of the GATT. These rules tend to supersede national laws and facilitate the privatization of biological knowledge and resources." Companies are now claiming to gain monopolies in the market (international market) over the products, which were earlier available freely for consumption purpose. For example, availability of seeds, plants and other organisms, now hold great value in the commercial market and, thus, has far reaching impact on biodiversity. This also leads to a great loss to the biodiversity. In addition, critics claim that "the privatization of genetic resources that have been engineered and patented accelerates the trend toward monoculture cropping." An example of this is visible in the Midwestern U. S. Corn Belt, where very few varieties of hybrid corn (hybrid) are available and cover millions of acres' land, thus supporting varieties of grasses, which were grazed for birds, butterflies, bees and other wild animals.

5.4.4 Technology

It is believed that science and technology is one of the most helpful and sorted tools to eradicate the problems related to biodiversity loss. But, this is not completely true. The scientific and technological developments have many adverse impacts on biodiversity. Moreover, the agricultural sector is assumed to be the most confusing one, when it comes to measuring the beneficiaries of technology. It is however highly unpredictable to figure out that whether the advancement in technology would act as a boon for this sector or not. But the recent development in this sector is of genetic engineering. "Genetic engineering has been cited as a driver of biodiversity loss." For example, In India, almost 50–60 thousand varieties of rice are cultivated by the poor farmers with the help of traditional farming practices.

It is because of the differences between soil and climatic conditions that help in growing different varieties of rice. "The GATT-TRIPs rules make it harder for farmers to harvest and reuse the seed of any rice variety that has been patented. Unlike hybrid species cultivated by plant breeders, genetically engineered plants do produce viable seed." Lack of access to seed stocks will cause the abandonment of much of India's biologically diverse agriculture, which in turn sustains healthy diversity in surrounding ecosystems.

A diagrammatic explanation of the types of economic failure is:

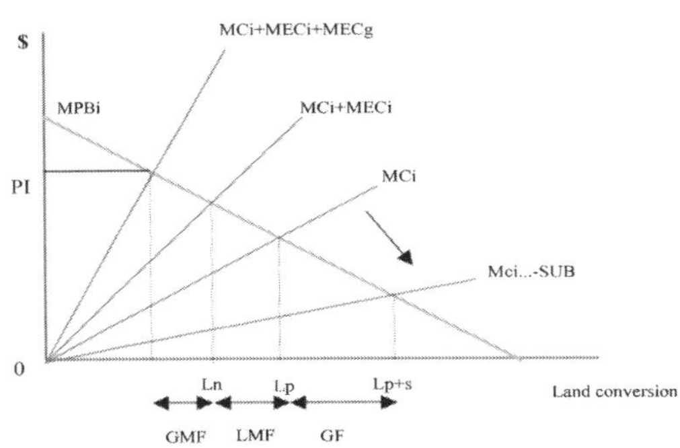

Figure 5.4: Measurement of economic failure.

Source: https://www.cbd.int/financial/values/g-economicvalue-iucn.pdf

The horizontal axis shows the land conversion, for example, land used for agricultural purpose. The vertical axis shows money to be spent on the conversion of land. The downward sloping line MPB_i is "the marginal private benefits" of land conversion, which elaborates the extra revenue collected by the farmers after they converted the land for agricultural purpose. The line MC_i is the marginal cost to the farmers after the land has been converted for agricultural practices. The `rational farmer' will equate MC_i and MPB_i to maximize profits. Hence, the amount of land conversion that actually takes place is L_P.

Now, if the farmers are provided with the subsidies for the conversion of land, then the effect of those induced subsidies is visible as a lowering of MC_i to $Mc_i - SUB$, where SUB is the subsidy provided to the farmers by the government. This illustrates that the private costs on the conversion of land have been lowered down. This, moreover, induces the landowner to expand the level of land conversion to L_{P+S}. The distance L_P to L_{P+S} is a measure of government failure (GF).

This would now change the level of land conversion. To find this, it is important to estimate the value of the two externality components: the local and global externality. This involves valuation. If the economists are aware of the value of damage made to the nation from land conversion, that is, losses incurred due to direct and indirect use values, then some estimate of the local external cost can be made. Figure 5.4 illustrates this loss as MEC_i where MEC_i is the marginal external cost imposed on the nation.

If this externality is "internalized," that is, if taxation and other higher cost bearing policies are imposed on farmers, then the land reserved for conservation would not be used for land conversion. This would help in maintaining an optimum level of land conversion, and the line optimum would shift towards L_n. Where, L_n is less than L_P, so that internalizing the externality involves less land conversion and hence more biodiversity conservation. The distance from L_n to L_P is a measure of the local market failure (LMF).

Furthermore, a similar approach could be used in case of global externalities, the value of the losses accepted by people outside the nation that owns the forest. The distance from L_G to L_n illustrates the global market failure situation or global appropriation failure (GMF). This analysis also provides with a rule for the `proper pricing' of land.

This can be understood by the equation:

$P_L = MC_i + MEC_i + MEC_g$

In other words, if the land is priced at the cost of conversion including the external cost of conversion, then there would be an optimal amount of land conversion. The optimal amount of land conversion is never zero. Although deforestation does take place during land conversion procedure, this approach is unrealistic and also unsatisfactory from the point of view of many economists.

5.5 ROLE OF ECOSYSTEM SERVICES IN VARIOUS ECONOMIC SECTORS

Economists have developed many methodologies to analyze economic value of biodiversity and ecosystem. The main focus has been levied on ecosystem services, which interface with economic processes. The economic processes contribute tangibly to the production and value creation of services rendered by the ecosystem and biodiversity. The provision of ecological services will be assumed to sum up and will include the relevant value of ecosystems and biodiversity, so that the term of ecological services will be used as shorthand for the value of ecosystems and biodiversity. Ecosystem services are considered as one of the several inputs that are necessary for the production of services through labor, technology, and capital. All the ecological services are directly or indirectly affected by the practices adopted for the production. The same has been described in the figure shown below.

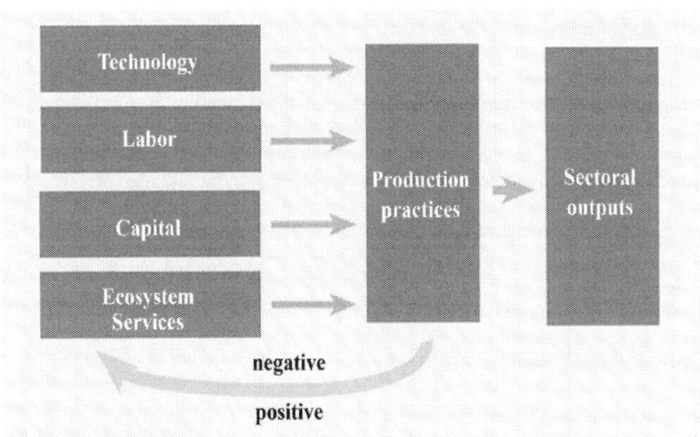

Figure 5.5: Relations among ecosystem services, other inputs, production practices, and sectoral outputs.

Source: http://www.undp.org/content/dam/undp/library/Environment%20and%20Energy/biodiversity/Report_ENG.pdf

Earlier these ecosystem services were abundant, and it was possible to degrade them without any noticeable effect on the sectoral outputs, as there were chances of either replacing those services or substituting them with any other service. However, there were many things like, when soil losses it fertility then there were chances of regaining the fertility through the use of fertilizers. As discussed above in section 9.2, Fisheries are a sector where it is not an easy task to replace the ecological services related to the fishery. Even aquaculture depends on wild capture fisheries for feed. Examples of costly technological input-substitutes for ecological services are:

- An increased water treatment structure and sediment removal machinery when the water quality gets degraded due to overexploitation of fishery.
- Another service that should be adopted by the humans is the injection of fertilizers and other related products to improve soil fertility.
- When there is a reduction in natural pest control, then it is necessary to increase the use of pesticide,
- Crop variation, and management effort;
- Reduced natural habitat for pollinator species requires artificial propagation and transport of honeybees or other pollinators; and
- Reduced natural habitat and biodiversity require alternative attractions for tourism.

The overall cost for an economic sector of degrading the ecosystem services depends on the capacity for substitution by other inputs and the costs of this substitution. The costs of technological fixes will change over time as ecosystem services degrade and require more inputs. An example is the need for intensive land reclamation and forest restoration given the impacts of mining. Moreover, there are limits on the extent to which human-made ecosystems through restoration try to substitute natural ecosystem services; but some of the ecosystem services cannot be fully replaced. As ecosystem degrades, the cost of substitution tends to rise. Thus, once an ecosystem service stops completely, no amount of substitute may work, effectively terminating the economic activity reliant on this loss ecosystem service.

The relationship between economies, biodiversity, ecosystem services, and natural capital are diverse. At one point of time, the use of land and natural resources by the commercial sector is regarded as the contribution of the sector to a biodiversity loss (economic drivers), whereas, at another point in time, commercial sector is highly dependent on the proper functioning of the ecosystem and the availability of ecological resources. The commercial sector uses the ecosystem services in such innovative manner that it leads to a new business opportunity for them. Consequently, while several business sectors are known to contribute to the degradation of ecosystems and biodiversity loss, these sectors can also play a proactive role in addressing the problem through their corporate environmental and social responsibilities.

5.5.1 Prevention of Negative Impact on Biodiversity

It is important for the commercial sector to take responsibilities to prevent the negative impact of commercial use on biodiversity. In many regions specially in the Arctic region, many current and emerging business opportunities, such as commercial fishing, tourism, and mineral and gas extraction, are creating a negative impact on biodiversity. The negative impacts caused by these commercial uses of ecological services are pollution and spills, habitat loss and fragmentation, the introduction of invasive species, and increases in disturbance to wildlife and people dependent on Arctic nature.

On the other hand, the risks are now increasing at an alarming rate due to the high concentration of commercial sector into mining, oil and gas development and shipping in the Arctic regions. Consequently, it is now high time for these sectors to realize and recognize that how important it is to manage the ecosystem in a sustainable manner. They should also take the benefits and values of ecosystem services to the local and indigenous communities, for example, provision of livelihood and spiritual importance of ecosystem services. However, if the importance of nature to various stakeholders is better understood, this would support better decision-making towards biodiversity that builds on the understanding of alternative outcomes and implications and adequately considers trade-offs and synergies between various services and related benefits. For example, consideration of ecosystem services can help to assess possible negative impacts occurring offsite, such as impacts of oil, gas and mining operations on fisheries and water purification.

5.5.2 Identification of Interdependencies and Opportunities

The commercial sectors are highly dependent on the supply of natural resources like timber, fish, and so forth. Similarly, businesses related to recreation and tourism rely on their access to nature. Furthermore, several industry sectors, such as mining and oil & gas, depend on ecosystems' ability to maintain water supply and mitigate flooding, erosion, and natural hazards at the locations of their operations. Moreover, the understanding of ecosystem services helps to highlight the often-overlooked dependencies that business sectors have on the environment, therefore helping to prevent the degradation of natural foundations that businesses depend upon.

Increased understanding of ecosystem services can also create opportunities for the development of sustainable businesses. For example, solutions for water management to gain best practices for water retention and purification. Another example for the identification of better opportunities in the commercial sector is, conservation and restoration of wetlands have been shown to be a potentially cost-effective option for managing water resources around the world. Furthermore, Arctic nature provides a range of opportunities for the development of novel value-added products. Organisms in the Arctic regions have evolved under extreme conditions, developing a variety of unique physiological and biochemical characteristics. All these traits of commercialization in ecological services provide a basis for a range of biotechnological innovations and related businesses with several companies estimated to be involved in research, development and/or sale of products derived from or based on the genetic resources of the Arctic and other countries.

5.5.3 An Initiative Made towards the Assessment of Business Ecosystem

There are many countries and regions where the responsibilities, risks, and opportunities associated with biodiversity and ecosystem services are still often overlooked and underestimated by the businessmen and are not fully accounted for along the entire value chain. To address this, the World Business Council of Sustainable Development (WBCSD) has developed guidelines for Corporate Ecosystem Review (CER) and Corporate Ecosystem Valuation (CEV) that help to improve corporate decision-making through assessing and valuing ecosystem services. "The aim of these guides is to help managers proactively develop strategies to manage business risks and opportunities arising from their company's dependence and impact on ecosystems."

The assessment of ecosystem services helps in strengthening the business performance in many ways. Likewise, the assessment of ecosystem services would not only help in forecasting the revenues for the business and reducing cost, but it would also help in determining various levels of liability and compensation so that it can also help in achieving social benefits for the community. In general, systematic integration of ecosystem service considerations into business decision-making supports the objectives for the long-term environmental and socioeconomic sustainability of the Arctic region, bringing benefits to both biodiversity conservation and people.

5.6 APPLICATION OF ENVIRONMENTAL POLICIES AND ECONOMICS

From the above discussion, we can conclude that the economic significance of biodiversity is both positive as well as negative. The economic drivers are harming the biodiversity and are leading to biodiversity loss. There are many commercial sectors that are also contributing to sustaining and conserving the biodiversity and the ecological services. It is also possible to measure the economic value of biodiversity through total economic value mechanism. This would help the ecologists to take steps in the conservation of biodiversity and would help the human beings sustain their lives in a better manner.

On the other hand, the role of government policies is significant in pushing forward the production of global biofuel as a crucial factor that helps in understanding the incentives that inspire biodiversity losses that are caused by the production of biofuels. In few instances, it is considered that biofuels are not economically competitive without the support of people. Some of the largest producers of biofuel across the world, including Brazil, the European Union, and the United States, are using a strong regulatory framework to sponsor biofuels. It is quite evident that the global production of biofuel is basically driven by governmental policy initiatives. In the present economic environment, the global regulatory framework is catching up with the possible shortcomings (social, environmental and economic) of the production of biofuel. Therefore, all the nations are aiming at initiating sustainable development criteria that are focused on eliminating negative impacts of biofuel production. Some examples of governmental initiatives that are encouraging biofuel production globally are illustrated below:

- European Biofuels Policies:

The fundamental policy drivers of production of biofuel in Europe are mentioned below:

- A community initiative namely "The European Biofuels Technology Platform" (EBTP) was formulated in the month of June in 2006. This initiative is considered as the centralized policy platform that helps in understanding the coherent EU biofuel policy. The EBTP was established with the aim of helping in the expansion of cost-competitive biofuel technology, that further helps in the establishment of a European biofuels industry.
- After the establishment of the 2003 Directive (2003/30/EC) on biofuels, the European Union (EU) members were committed to include biofuels in their transportation system at the rates of 5.75 and 10% by 2010 and 2020, respectively. European Union members are given flexibility for implementing the 2003 Directive meanwhile only the goals are stated in the directive. The goal that was set for 2005 was 2%, which was missed by 0.6 points, while the goal for 2010 did not seem to get accomplished for the EU as a whole.
- Another Directive namely "The Energy Taxation Directive" (2003/96/EC) allows the member of European Union to allow discounts and exclusions of biofuels in transportation system to help in promoting their incorporation into the market for automobile transport.
- Some member states of European Union are announcing biofuel blending levels as obligation. Such announcement obliges fuel suppliers to use a percentage of biofuels (either ethanol or diesel) in the market.
- The Energy Crop Scheme (ECS) which was launched during the CAP reform of 2003, provides support in the development of energy crops. The ECS scheme replaced the scheme of Non-Food Set-Aside (NFSA) which is considered as a CAP transfer scheme that provides payment to the farmers who were producing non-food. However, under the new ECS scheme, covering a fixed region of 2 million hectares including the new EU member states (it was 1.5 million before 2007), an aid of €45 per hectare is available. If the new scheme is fully implemented, this scheme will cost €90 million to the European Union.

However, it is estimated that the global land demand will increase by 10% by 2020 agricultural items such as for wheat, maize, oilseeds and sugar cane. This increase will occur even in a baseline where yield improvements of biofuel policies are assumed. It is also noted that with additional policies of biofuels in the United States and the EU, the growth of 5% may be anticipated by 2020. It is further estimated by The Institute for Prospective Technological Studies (IPTS) that between 12 and 15% of the arable land of European Union would be required to meet the goal of European Union Biofuel Directive.

- Policies for Biofuel in the United States of America:

Several federal agencies are involved with the United States, which are committed to the promotion of biofuels. Some of these federal agencies include the Department of Energy and the Department of Agriculture. The Secretary of Agriculture launched the Bio-based Products and Bioenergy Coordination Council (BBCC) in 1995. This council was established to make a platform through which different activities can be promoted and coordinated by USDA agencies. These activities might include exploration, expansion, transmission of technology, commercialization, and promotion of bio-based products and bioenergy using renewable domestic materials of agriculture and forests. This concept is similar to the European Biofuel Technology Platform scheme.

It is researched that several provisions were included in The Farm Bill passed by the US Congress in 2002. This Bill was passed to grant public support to the agricultural sector and push the people for using biofuel production in the United States of America. The biofuel production includes support for biofuel goods, biorefinery equipment, and bio-based products.

Moreover, the government of the United States of America initiated a policy namely "The Energy Policy Act" in 2005 which is considered as a large energy policy package. This policy act comprises various provisions that are related to biofuels and bio-based production. The main drivers of public policy in the US that are confined in the Energy Policy Act of 2005 that initiates the production of biofuel are described below:

- Section 932. In this section, the Bioenergy Program is said to create a linked between the bioproducts, industrial and academic institutions, and Department of Energy's biomass to boost the progress of biofuels, bioproducts, and biorefineries.
- Section 942. In this section, information related to the Production Incentives for Cellulosic Biofuels is contained. This section

provides incentives to confirm that by 2015, the annual production of one billion gallons of cellulosic biofuels is accomplished.

- Section 1501. This section determines Renewable Fuels Standard. This section determines that the gasoline sold in the United States of America should be blended with ethanol. It is revealed that 4 billion gallons of ethanol were mixed with gasoline in 2006, and this requirement further amplified annually to 7.5 billion gallons by 2012. It is further estimated that the essential capacity of renewable fuel will increase to a minimum of 250 million gallons of cellulosic ethanol for 2013 and beyond.
- Biofuel policies in Brazil:

It is considered that Brazil is the largest producer and exporter of bio-ethanol across the world, with a total manufacturing level of approximately 25 billion liters in 2008. The national alcohol program (PROÁLCOOL) for bio-ethanol in Brazil was initiated in 1970, which was established with the aim of accomplishing independence in terms of energy and exploitation of the large arable land surface of the nation to produce sugar cane for the purpose of producing ethanol and bioethanol.

REFERENCES

1. Chivian, E., & Bernstein, A. (Eds.). (2008) *Sustaining life: How human health depends on biodiversity* (Illus. ed.). New York: Oxford University Press.
2. ConservationTools. (2017) *Economic benefits of biodiversity: Conservation tools*. Retrieved October 14, 2017, from http://conservationtools.org/guides/95-economic-benefits-of-biodiversity
3. Costanza, R. d., Groot, R. D., Farber, S., Grasso, M., Hannon, B. (1997). The value of the world's ecosystem services and natural capital. *Nature, 387*(6630), 253-260.
4. Costanza, R. d., Groot, R. D., Sutton, P., Ploeg, S. v. d., Anderson, S. J. (2014). Changes in the global value of ecosystem services. *Global Environmental Change, 26*, 152-158.
5. Groot, D., Wilson, M. A., Boumans, R. M. J. (2002). A typology for the classification, description and valuation of ecosystem functions, goods and services. *Ecological Economics, 41*(3), 393-408.
6. Kettunen, M. (2015). *Business and ecosystem services* [Blog]. Retrieved October 14, 2017, from http://arctic.blogs.panda.org/default/business-and-ecosystem-services/
7. Nrcan.gc.ca. (2017). *How does the forest industry contribute to the economy?|Natural Resources Canada*. Retrieved October 14, 2017, from http://www.nrcan.gc.ca/forests/report/economy/16517
8. Pearce, D., & Moran, D. (1994). *The economic value of biodiversity IUCN—the world conservation union*. Retrieved October 14, 2017, from https://www.cbd.int/financial/values/g-economicvalue-iucn.pdf
9. Pisupati, B. (2017). *Economic value of biodiversity* [The Hindu Business Line]. Retrieved October 14, 2017, from http://www.thehindubusinessline.com/opinion/economic-value-of-biodiversity/article3297054.ece
10. UNDP. (2010). *Importance of biodiversity and ecosystems in economic growth and equity in Latin America and the Caribbean: An economic valuation of ecosystems*. Retrieved October 14, 2017, from http://www.undp.org/content/dam/undp/library/Environment%20and%20Energy/biodiversity/Report_ENG.pdf

6
CHAPTER

CONSERVATION BIOGEOGRAPHY

"People need to look at wildlife conservation in its totality. As soon as you lose the apex predator, it has harmful consequences right down the food chain."

—Thomas Kaplan

CONTENTS

ABSTRACT

In this chapter, we will talk about the ways to conserve the ecosystems, animals and plants that are on danger of extinction. Also, we will briefly discuss about the historic extinction of certain group of animals, the time of extinction, scope of extinctions, and also the effects of these extinctions on global species diversity. In the second half of the chapter, biogeography and evolution are discussed, as well as the extinction crisis and the causes of extinction are well structured. At the end of the chapter, the name and reason behind the extinction of top 10 animals are mentioned.

6.1 INTRODUCTION

Conservation biogeography is a comparatively new hypothetical attempt that brings conservation as well as practical apprehensions to the forefront by merging the traditions of biogeography with the concerns of conservation biology. To examine the spatial organization of biological diversity, biogeography is a well-established scientific discipline. Classically, this is done by:

- Observation of biophysical factors as well as ecological factors that mark the distributions of species and of biotic communities or ecosystems;
- Inspection of evolutionary mechanisms and antique events that altered those species as well as their distributions through plate tectonics, species interactions, climate change, or other events or processes.

Often, biological data to explicate the evolutionary lineages of a group of organisms are used by a biogeographer, which then to understand the histories of the relevant parts of the earth's surface questions the consequences of those lineages. Questions are posed by the biogeographers about the distributions of assemblies of species, including the ecosystem processes that structure the land cover of the earth as well as the types of vegetation they depend on. "Conservation biogeography uses both these paradigms to examine the implications of past and present distributions to plan for the protection of biodiversity under current and possible future conditions." However, given its link to Comparative biology, Biodiversity Conservation and Conservation Biology we can study altogether. Indeed, there are several Journals such as Biodiversity and Conservation, Conservation Biology and

Conservation Biogeography. The last one from Wiley deals with diversity and distributions Conservation biogeography and involves the application of biogeographical principles, theories, and analyses to problems concerned with the distributional dynamics of taxa individually and collectively, to problems concerning the conservation of biodiversity'. It is considered as a subdiscipline of both Biogeography and Conservation Biology.

6.1.1 What is Conservation Biology

In the 1970s, when biologists began to describe the current scenario as a "biodiversity crisis, conservation Biology came into existence as a distinct discipline. Death rates were seen to match those during megafaunal extinctions, for example, the dinosaurs during the Cretaceous. In 1987, "The journal Conservation Biology" was created and in 1985, "The Society for Conservation Biology" was founded.

Conservation biology ascended from two opposed views of nature, which were:

- "The Romantic-Transcendental Conservation Ethic—from the writings of Ralph Waldo Emerson, Henry David Thoreau, and John Muir. Nature viewed in a quasi-religious sense and as way to cleanse the human soul. The tradition continues today in the Sierra Club."
- "The Resource Conservation Ethic—from Gifford Pinchot, this is the basis of much of the activities of government agencies like the Forest Service. It is anthropocentric, and views nature as resources to be used wisely."

A third view was joined by the above two opposing views. This view prevails in Conservation Biology today. It is called "The *Evolutionary-Ecological Land Ethic*—Aldo Leopold is its source, clearly presented in *A Sand County Almanac* (1949)." However, unlike Leopold, modern ecology/ conservation biology views ecosystems in a non-equilibrium context.

"Conservation biology is the scientific study of how to maintain diversity with increasing human population size and development." Conservation biology is a synthetic science, which is built from biogeography, ecology, economics, philosophy, population biology, population genetics, and anthropology, with the purpose of emerging ideologies as well as approaches to preserve diversity. Conservation biology's main objectives are:

- To explore human impacts on global diversity;
- To advance applied methods to indorse human development without extinguishing biological diversity.

The conservation biology focuses on three guiding principles which are:

- Principle 1: Evolution is the basic axiom that unites all of biology. (The evolutionary play);
- Principle 2: The ecological world is dynamic and largely non-equilibrial. (The ecological theatre);
- Principle 3: Human presence must be included in conservation planning. (Humans are part of the play).

In 1994, Primark proposed another view of the basic principles from a different introductory text, which is:

- "Diversity of organisms is good" Humans usually appreciate and value biodiversity. Human-mediated extinction of populations and species is bad.
- "Ecological complexity is good and in many instances mandatory to species survival." Harm to one species may result in 'cascading' effects on others.
- "Evolution is good."
- "Biological diversity has intrinsic value."

Two major developments have influenced the development of biogeography during the 20th century; these are "the recognition of continental drift" and "the development of a quantitative theory of island biogeography." In 1912, Wegener first described the continental drift, the acceptance of which took quite a while. In 1957, Philip Darlington published a monograph "—that did not accept continental drift and "fanciful" explanations were created for relations among the southern continents. A revision about a decade later was published by him which used continental drift explanations. Last, the quantitative theory of biogeography, traceable to MacArthur and Wilson comes up. It is from quantitative biogeography that conservation biology arose.

6.2 HISTORICAL EXTINCTIONS

In the recent literature to describe historical extinction episodes, the terms like "mass extinction," "mass killing," "mass murder" and "catastrophic event" have all been used. The use of such terms suggests indication of unexpected disasters that have been spread out in substantial portions of the earth's plant as well as species. They give the impression of an overwhelmed globe littered with the remnants of dead organisms. For instance, in 1983, Allaby and Lovelock said that 65 Ma ago the earth had collided with a small planet, in their book entitled "The Great Extinction." They defined thc effect as a conflagration when "volcanoes erupted, tidal waves swept the oceans, and earthquakes shook the continents. For years, a dust cloud shrouded the earth blocking out the sun. Three quarters of all species died including the mighty dinosaurs." Finally, in 1990, A1-varez and Asaro in their Scientific American article mentioned a "sensational crime" which almost killed off half of all the life existing on earth and provided evidence which supposedly showed that a giant asteroid or comet had committed this "mass murder."

There are many imperative queries about historical extinctions that need to be discussed:

- "How did the mass extinction idea get started?"
- "Did the events occur so suddenly that they need to be called mass extinctions?"
- "Were they so widespread that they deserve to be termed global events?"
- "Did they really have drastic effects on the world's species diversity?"
- "Did they have a common cause?"
- "Did they have important biogeographic and evolutionary effects?"

6.2.1 Tempo of the Extinctions

In 1990, Briggs helped in obtaining the recent information on the time span over which each of the major extinctions occurred. The major time of mass extinction occurred in the middle part of the Vendian, about 650 Ma ago, so slight data is available about the timing or severity that its recognition may be early. Although extensive evolutionary and community changes did take place, still the end of the early Cambrian and the end of the Cambrian are sometimes termed extinction events, but the evidence is not clear cut. During the Ashgillian Stage near the end of the Ordovician, the first widely recognized extinction episode took place. The measure has been estimated at 1–2 Ma (Brenchley, 1989, 1990) or "several million years or so" (Boucot, 1990b). In 1988, McGhee analyzed the general marine-invertebrate data on the Frasnian-Famennian extinction, towards the end of the Devonian. A stepwise sequence of extinctions covering a period of more than 1 Ma was found by Schindler in 1990.

The major event that was examined so far was the Cretaceous-Tertiary (K/T) extinction. As far as time is concerned, it seems that some large terrestrial animals, like the pterodactyls and the dinosaurs had been decreasing for several million years former to the K/T boundary. Smaller land animals appear to have been barely affected. In North America, extensive destruction of higher plants took place which was not seen anywhere else. Over about 2.5–2.7 Ma, most of the marine benthic animals that were affected vanished. Over about 0.8–1.0 Ma, some planktonic forms also declined.

Furthermore, in 1990, McLaren and Goodfellow stated that "Impacts by giant, extraterrestrial bodies are still being proposed as the primary cause for almost all of the extinction episodes." The tempo of the extinctions is inconsistent with hypotheses involving single high-energy events, even if the impacts and volcanic eruptions may have occurred during or near times of biotic decrease. In response to environmental changes, these were gradual deteriorations in organic diversity, which were developed. When applied to extinction episodes in the geologic record, terms such as "mass extinction," "mass killing," or "catastrophic event" are majorly misleading.

6.2.2 Scope of the Extinctions

In the Phanerozoic history of the seven episodes, which have been identified as times of great extinctions, the one that truly deserves considered to be a global phenomenon is the Permian/Triassic event. Drastic decrease in diversity was seen in both marine and terrestrial environments in the Permian/Triassic event. As illustrated by briggs in 1990, "For the K/T boundary, which has often been described as a global catastrophe, the overall effect on the terrestrial biota was small compared to the reduction of the tropical marine life." In the later observation, it was seen that many vertebrates and invertebrates' groups survived quite well, but the population of remaining dinosaurs eventually died off. A temporary setback in biomass was received by the angiosperm plants, which had no visible effect on their enduring growing diversity.

6.2.3 Effects on Global Species Diversity

Due to the dependence on fossils from the shallow waters of the marine tropics, estimates of the severity of so-called global extinctions have often been biased. During the times of environmental crisis, a better survival is always shown by the samples from high latitudes or the deep sea. Another problem is that the fossil record is not totally dependent to the extent that it can be used for a precise valuation of the extent of extinction measures. However, it is expected that during the coming years it is necessary to have the information about the magnitude of widespread extinctions from detailed studies of discrete assemblies where the pasts of distinct phyletic lines have been resolute. Finally, there are signs that the effects of extensive destructions on the global species diversity have been exaggerated. For example, "estimates of species extinguished at the K/T boundary range from 50 to 80% of all species."

6.3 BIOGEOGRAPHY AND EVOLUTION

In recent years, there has been significant speculation about the evolutionary possessions of extinction incidents which gives a reliable message of the impression that a "wiping out of the old forms to make way for the new" has lasting evolutionary welfares. In 1984, Gould suggested that "extinctions might be the primary and indispensable seed of major changes and shifts in life's history; that if environments did not undergo such changes, evolution might well grind to a virtual halt." In 1986, during an extinction event, Raup referred to the common belief that the species which can adapt themselves in the changing environment in a better way than others would survive, and this ultimately would lead to an enhancement in the whole biota. In 1987, Eldredge expressed the firm belief that, "without extinctions to free up the ecological niches, life would still be confined to a primitive state somewhere on the sea bottom." On the other hand, Stanley (1987) stated, "Had the dinosaurs survived, there is no question that we would not walk the earth today. Mammals would still remain small and unobtrusive, not unlike the rodents of the modern world." In 1986, Hsti developed a new theory of evolution in which the idea behind the major extinctions has been mentioned. In this "new catastrophism," evolutionary advances would take place as the survivors acclimatize to places formed by destruction events. In his view, "it is time to awaken to the absurdity of the idea of natural selection." the common assumption of evolutionary benefits is not supported by the available data on the widespread extinction patterns.

A number of authors have observed that, "the warm-water or tropical marine organisms have been more prone to extinction during crises than those in cold or deep waters and noted that this pattern has been found in late-Cambrian trilobites, late-Ordovician and early-Silurian brachiopods, late-Devonian brachiopods and corals, and late Cretaceous gastropods and planktonic foraminiferans." In regard to all large extinctions, in 1986 Jablonski stated those low-latitude taxa, particularly in reef communities, were invariably more severely affected than cosmopolitan taxa, polar, and temperate. In regard to the K/T crisis, in 1986 Gerstel et al. found that cool-adapted species of planktonic foraminiferans migrated to low latitudes substituting warm-adapted species, which become extinct in the particular area due to some climatic change or because they could modify themselves according to the present environmental needs.

It could be said, however, that for some groups of marine animals that have good fossil records, like the bivalve mollusks, the benthic Foraminifera, and the herma typical corals, the younger and the more advanced genera are found in the tropics while the more ancient ones are found on the higher latitudes. In other groups, the higher-latitude representatives have a simpler, and more primitive morphology, as the fossil history is not as well known. Moreover, recent systematic work has shown that the youngest and most specialized genera and species are occupied by the tropics and especially certain centers of origin within them. The patterns of increasing age with increasing latitude are applicable for all the taxa. As the taxa slowly disperse to higher latitudes, they become older.

On the other hand, the younger and more advanced species continue to progress in the tropics. Most drastic effect on the youngest, most advanced taxa is seen when a widespread tropical or circumtropical extinction occurs. It is clear that most or almost all of the great extinctions specially eradicated the epipelagic organisms and the tropical shelf. This created an ecological vacuum which progressively reoccupies simpler, eurythermic organisms from higher latitudes, maybe from deeper water and other refuges. This course of the sequence by older and more primitive forms may set back the evolutionary clock to an earlier time.

Furthermore, the production of successful new species and genera has been interrupted by the damage to the tropical centers of origin. In 1985, Kitchell and Carr supported the concept of evolutionary setbacks by their model, which predicts that "extinction episodes generally delay taxonomic turnover, rather than being precipitated by them."

In 1980, Gould and Calloway gave the example of the brachiopods versus the pelecypods as part of their argument for the evolutionary advantage of extinctions. According to their observation pelecypods gained the advantage as the brachiopods were hit harder by the P/T extinction, and this ultimately gives a clear picture that the pelecypods are obviously the superior competitors than brachiopods. Pelecypods instead of being attached to the substrate, they can actively move about and pump water through their gills far more efficiently, as well as can grow faster. The present-day supremacy of pelecypods compared to brachiopods in the warm, shallow habitats of the world, is a tribute to the superior competitive ability of the former, as brachiopods are mainly relegated to colder and deeper waters. In other groups, also similar evolutionary replacements take place.

In the mid-Paleozoic period, the gymnosperms began their development for about 100 Ma, as well as along with the pteridophytes they have increased their diversity. gymnosperms became somewhat the more diverse in the latter part of the Paleozoic. Both gymnosperm and pteridophytes have suffered equally during the extinction, but still, the gymnosperms continued their expansion while the pteridophytes gradually played a minor role. At the beginning of the Cretaceous, the angiosperms arose, and rapidly they established into the leading plant group. In each case, the more primitive group that was prospered sustained to do well for about 100 Ma beyond the time of origin of its successor. Both pteridophytes and gymnosperms ultimately started to decline, but these changes had become apparent prior to the K/T and P/T boundaries, respectively.

Despite a brief interruption at the K/T boundary, the angiosperm increased continuously. As noted by researchers that "The biogeographic consequences of the changes in the major plant groups are that the angiosperms now dominate the tropical and warm-temperate zones of the terrestrial world where edaphic and rainfall conditions are favorable." In contrast, the gymnosperms predominate in less favorable locations and at higher latitudes, while the pteridophytes tend to be scattered in marginal habitats. The large-scale extinction events are neither feature the origination of the major plant groups nor the relative decline of two of them. It is still believed that the evolutionary success of the mammals was due to the extinction of the large reptiles at the K/T boundary. Of greater importance is the fact that mammalian evolution did make considerable progress in the late Cretaceous when the large dinosaurs were still prevalent.

There existed five genera of marsupials, seven genera of placentals, and ten genera of multituberculates in northwestern North America, the above data provide evidence that "mammalian radiation was already well underway before the large dinosaurs left the scene." In 1986, Sloan et al.

described fossils from the Hell Creek Formation, which suggested that dinosaur extinction, as well as the ungulate radiation, were concurrent events, which implies that competition exists in that region. In the late Jurassic, feathered dinosaurs (birds) came on the scene, but in diminishing diversity, the older pterosaurs were able to carry on for another 70–80 Ma. Mesozoic amphichelyd turtles is another example concerning the above. Although the worldwide replacement process is a bit slow, still in North America, the replacement process seems to have been somewhat accelerated across the K/T boundary. Amphichelyds finally became extinct in the late Pleistocene. The replacement by the two successor groups took about 150 Ma. These exceptionally slow, evolutionary replacements of chief groups seem to be the practically constant rule in both marine and terrestrial environments. In the terrestrial environment, "when one compares the biota of the high-latitude versus the low-latitude parts of the world, it becomes at once apparent that the former usually contains the older and more primitive organisms.

"Higher-latitude forms are better adapted to extremes of humidity as well as varying temperatures. An invasion of more primitive forms from higher latitudes is expected from the latitudes as a higher rate of extinction occurs in the tropics. The conclusion to this is to set an end clock to evolutionary change on both sea and land. At the present state, it appears valuable to theorize that the great antique destructions were disasters that delayed rather than improved the course of evolution. In 1987, as Vrba has settled for African mammals, "It is the specialists that are shaved off wholesale from the Tree of Life by mass extinction. The system is reset, leaving a few hardy generalists to start all over again." There is a catastrophic extinction event occurring in contemporary time, although the fossil evidence does not support the concept of historical mass extinctions or mass killings.

In 1990, Raven estimated that "by the first quarter of the next century, the world will have lost two million out of a minimal world total of 10 million animal species and about 65 000 out of 300 000 species of vascular plants." These losses are occurring rapidly, due to habitat destruction by humans, that is unparalleled in Phanerozoic time. Historic extinction episodes were so steady that many lineages were able to accommodate in an ecological as well as evolutionary sense. The tempo of the current extinctions prevents any such adjustments.

6.4 THE EXTINCTION CRISIS

It is terrifying but true: "Our planet is now in the midst of its sixth mass extinction of plants and animals—the sixth wave of extinctions in the past half-billion years." Since the loss of the dinosaurs over 65 million years ago, the worst spate of species die-offs was experienced. Extinction is a natural phenomenon which is occurring at a very frequent pace that is every year at least one to five species extinct, and if this continues like this, then sooner will be the day when all the species will get vanished. According to scientists estimates approximately 1000–10000 species are lost every year, with literally dozens going extinct every day. With as many as 30–50% of all species possibly heading toward extinction by mid-century, it is surely going to be a scary future certainly.

The current crisis is almost entirely caused by humans, unlike past mass extinctions, caused by events like volcanic eruptions, asteroid strikes, and natural climate shifts. Almost 99% of presently vulnerable species are at danger from human activities, principally those driving habitat loss, and global warming. Because every species' extinction potentially leads to the extinction of others bound to that species in a complex ecological web and because the rate of change in our biosphere is increasing, numbers of extinctions are likely to snowball in the coming decades as ecosystems unravel. Species diversity confirms ecosystem flexibility, giving ecological communities the room they want to bear stress. A comprehensive strategy for saving biodiversity must also include habitat types with fewer species, like grasslands, tundra, and polar seas for which any loss could be irreversibly devastating, while conservationists often justifiably focus their efforts on species-rich ecosystems like rainforests and coral reefs which have a lot to lose.

Most of the biodiversity's benefits take place at a local level while much concern over extinction focuses on globally lost species, and conserving local populations is the only way to ensure genetic diversity critical for a species' long-term survival. Approximately 1,000 species have gone extinct in the past 500 years; these include the woodland bison of West Virginia and Arizona's Merriam's elk to the Rocky Mountain grasshopper, passenger pigeon, and Puerto Rico's Culebra parrot. But this is not the exact figure that was recorded, there were many more species which disappeared unexpectedly, and scientists still have no records for such species. Still, scientists have no exact number about the number of species that are in danger of becoming extinct. According to the estimate of David Wilcove, a conservation scientist, "there are 14,000–35,000 endangered species in the United States, which is 7–18% of U. S. flora and fauna."

The IUCN has evaluated roughly 3% of described species and identified 16,928 species worldwide as being threatened with extinction, or roughly 38% of those assessed. The IUCN reports that the world won't meet a goal of reversing the extinction trend toward species depletion by 2010 In its latest four-year endangered species assessment.

Thousands of species are at risk of disappearing forever in the coming decades, these include:

- **Amphibians**

  Amphibians, among all other groups of animals, have the highest rate of endangered species. According to the current estimates by scientists, roughly 6,300 known species of amphibians are on the verge of extinction. The current rate of extinction of amphibians ranges from 25,039 to 45,474 times the background extinction rate. Due to loss of habitat, water, and air pollution, ultraviolet light exposure, climate change, introduced exotic species, and disease the population of frogs, toads, and salamanders started disappearing. Endangered amphibians should be viewed as the canary in the global coal mine because of their sensitivity to environmental changes, which signals subtle yet radical ecosystem changes that could ultimately claim many other species, including humans.

- **Birds**

  Birds are often the most visible and familiar wildlife to people across the globe, and they occur in nearly every habitat on the planet. As such, they provide a significant bellwether for tracing changes to the biosphere. Due to the profound changes occurring on the planet the population of the bird started declining. The population of bird is falling almost in all habitats in response to the human activities. According to the 2009 US report on the state of birds, it was found that "251 (31%) of the 800 species in the country are of conservation concern."

Globally, BirdLife International estimates that "12% of known 9,865 bird species are now considered threatened, with 192 species, or 2%, facing an 'extremely high risk' of extinction in the wild, two more species than in 2008." Most of this decline is caused due to habitat loss and degradation, but the influences of hostile species as well as arrest by collectors play a big role, too.

- **Fish**

  It is not at all surprising to know that there are many fish species that are endangered in both freshwater and marine habitats after seeing the increasing demand for water, the dumping and accumulation of various pollutants, the damming of rivers throughout the world, and invasive species make aquatic ecosystems some of the most threatened on the planet.

  The American Fisheries Society found 700 species of freshwater or anadromous fish in North America as being imperiled, amounting to 39% of all such fish on the continent. In North American marine waters, at least 82 fish species are imperiled. In 2010, according to IUCN, "across the globe, 1,851 species of fish, 21% of all fish species evaluated, were deemed at risk of extinction, including more than a third of sharks and rays."

- **Invertebrates**

  There is no exact estimation to the number of invertebrate species present on earth, but according to the rough estimates, 97% of the overall species of animals on the earth are invertebrates. This group includes species from butterflies to mollusks to earthworms to corals. The IUCN has evaluated about 9,526 species out of the total of 1.3 million known species, almost 30% of the species evaluated at risk of extinction from this estimated number. Moreover, factors like water pollution, groundwater withdrawal, and water projects, severely threats the freshwater invertebrates, while due to deforestation, especially because of the rapid destruction of tropical rainforests, a large number of invertebrates of notable scientific significance have become either endangered or extinct. In the ocean, reef-building corals are declining at an alarming rate: "2008's first-ever comprehensive global assessment of these animals revealed that a third of reef-building corals are threatened."

- **Mammals**

  Possibly one of the most prominent essentials of the current extinction crisis is the fact that the majority of our closest relatives, the primates are severely threatened. About 90% of primates are disappearing at a very fast rate; primates are the group that includes monkeys, lemurs, lorids, galagos, tarsiers, and apes (as well as humans) live in tropical forests. According to the rough estimates of IUCN, almost 50% of the world's primate species are at risk of extinction. Overall, the IUCN evaluates that "half of the globe's 5,491 known mammals are declining in population and a fifth are clearly at risk of disappearing forever with no less than 1,131 mammals across the globe classified as endangered, threatened, or vulnerable." In addition to primates, marine mammals including some species of porpoises, dolphins, and whales are among those mammals slipping most quickly toward extinction.

- **Plants**

  Plants provide the oxygen we breathe and the food we eat and are thus the foundation of most life on Earth. This all is done through the process of photosynthesis. The majority of medicines that are available in the market today are made from the plant sources only. The IUCN has evaluated that "Of the more than 300,000 known species of plants, only 12,914 species, finding that about 68% of evaluated plant species are threatened with extinction." Plants are highly vulnerable to extinction as they cannot move readily from one plane to another if their habitat is lost. Global warming is expected to worsen this problem.

Already, scientists say, quick and dramatic changes in the range and distribution of plants around the world are caused by warming temperatures. With plants making up the base of the food chain and the backbone of ecosystems, that's very bad news for *all* species, which depend on plants for shelter, food, and survival.

- **Reptiles**

  As per the evaluation made by IUCN, "21% of the total evaluated reptiles in the world are considered endangered or vulnerable to extinction, 32 reptile species are at risk, about 9% of the total." With at least 28 island reptiles having died out since 1600, island reptile species have been dealt the hardest blow. Habitat destruction and the invasion of non-native species, which prey on reptiles and compete with them for habitat and food are the main threat to reptiles.

6.5 CAUSES OF EXTINCTION

Due to natural occurrences like climatic heating or cooling or changes in sea levels, animals have become extinct. Seeing the present scenario, human activities have also become the major cause of the extinction of many animal species. The modern extinction of animal and plant species are chiefly due to habitat destruction as farming land expands and forests are cut-down, along with pollution, the introduction of alien species, as well as over fishing or hunting.

When the population of organisms, normally species end in an area, the term is known as extinction. The death of the last individual of the group is generally considered as the extinction of particular group in the environment. Because a species' latent range may be very large, determining this moment is difficult and is usually done retrospectively. Lazarus taxon is the phenomenon which ultimately is talked about in concern to the above; in this phenomenon, a species alleged extinct brusquely re-appears after a period of apparent absence. Several influences are driving an unparalleled rate of extinction of animal as well as plant species globally. Although extinction is a natural process, the rate at which current extinction is taking place is clearly not, and all scientific evidence indicates that the activities of mankind are the primary engine behind most recent and present extinction events. Some of the chief causes of extinction are:

- **Habitat Loss**

Destructive change to landscapes or environments, either through human processes (such as construction, deforestation, changing land use for agriculture, artificial land drainage etc.), or natural phenomena (such as floods, volcanoes, hurricanes etc.), is the single utmost danger to the biodiversity of Planet Earth, and the extreme cause of extinction in the world. The animal population will become extinct, when a plant or an animal does not have a habitat, and cannot adapt to a different environment.

- **Unregulated or Illegal Killing, Hunting or Poaching**

A human cause of extinction that may represent a major or dominant factor in the decline of certain species is hunting and poaching rare plants and animals, predominantly those that have a small or slow-regenerating population overall or are endemic to a small geographic area.

Unfortunately, across the world, numerous socioeconomic issues drive hunting as well as poaching of threatened animal and plant species, and where this occurs at unsustainable, unregulated levels, susceptible species are pushed towards destruction. However, although at a national or international level (i.e., the Convention on International Trade in Endangered Species), regulations and legislation may exist, habitually adequate substructure, consciousness or capitals are in place for any active influence. Due to ignorance or misconceived stereotypes, sometimes many of endangered the plant and animal species are killed, for instance, the case of bats, snakes, and arachnids that are commonly, but incorrectly, perceived to be aggressive or necessarily dangerous.

- **Introduced Species**

"The introduction of plant and animal species that are not endemic to a given locality is both a natural and human process that often has disastrous knock-on consequences for local biota, often including extinction of native taxa." Through regular dispersal processes over short geographic distances (i.e., seed being blown in the wind, or in the crop of a bird), many species are introduced that are not native to a given area, but the incidence of such processes if happening naturally is low. Artificial or accidental introduction of non-native plant and animal species occurs much more commonly, particularly as people travel more widely and recurrently across the globe.

- **Pollution**

A natural or human cause of extinction is pollution that can take many forms. "Natural pollution events may result from cataclysmic geographic processes (volcanic eruptions, floods, earthquakes, etc.), or from over-population of ecosystems by specific species (red tide) or other processes." Local extinction events are commonly caused by natural pollution events, but infrequently are adequately varied scale to cause complete extinction of substantial records of plant and animal taxa. Human pollution usually arises when toxic substances are dumped, either advertently or inadvertently, into biologically diverse areas of our planet. Anthropogenic pollution may have knock-on significances, for instance, eutrophication. the complete extinction of plant and animal taxa, particularly those that are endemic to a small geographic area, or have a small or slow-regenerating population overall are caused by the large-scale anthropogenic pollution events (i.e., oil spills). Moreover, due to pollution, the entire ecosystem gets affected, including humans, for instance, the pesticide DDT, which was used until the 1970s against arthropods species, but causes catastrophic impacts at all ecological levels, from the water and soil, through water feeders, ground arthropods, predators, and humans.

- **Competition**

Competition drives ongoing evolutionary processes, and plant and animal taxa that are unable to adapt may be out competed and naturally displaced from their habitat, and pushed toward extinction, over prolonged periods of time.

- **Disease**

The spread of disease may be both a natural and human factor behind extinction.

Naturally occurring diseases that trouble exact animal or plant taxa may be unintentionally spread by humans with catastrophic costs, for instance, "Dutch elm disease, which is a fungal disease of elm trees spread by the elm bark beetle." It was originally believed that the disease is native to Asia only, but later it was found that the disease has been accidentally introduced into North America and Europe, where it has confounded natural populations of elms which had not had the prospect to evolve resistance. Numerous other issues contribute to the extinction of animal and plant taxa of Planet Earth.

6.5.1 Top 10 Extinct Animals

West African Black Rhinoceros

In the southeast regions of Africa, the Black rhinoceros are majorly found. The measurements of these rhinoceros were approximately 1.4–1.7 m in height, 3–3.8 m long and 800–1300 kg in weight. These rhinoceros' species had two horns, one measuring 2–55 cm and the other between 0.5–1.3 m. Their diet comprises shoots and leafy. Some native people also believed their horns had medicinal properties, which was never proven by any scientific facts. However, preservation action was taken to protect the species during the 1930s, but still, the numbers were constantly declining. It was in 2006 when the last West African Black Rhino was seen in Cameroon. Black rhinoceros were officially declared as extinct species by the end of 2011.

Figure 6.1: West African black rhinoceros.

Source: https://onekindplanet.org/top-10/top-10-worlds-extinct-animals/

Baiji White Dolphin

Baiji White Dolphin, also called the Chinese River Dolphin, can only be found in the Yangtze River in China. These mammals weigh up to a quarter of a ton and could grow to eight feet long. Due to their tiny eyes and very poor eyesight, these mammals rely on echolocation to navigate and hunt for pray. For over 20 million years, they lived in the Yangtze; after 1950, their numbers declined extremely. As China industrialized, the river was used for fishing, transportation, and hydroelectricity due to which it had a massive effect on the mammals, and they suffered a lot due to industrialization. Although white dolphins are not officially recorded as extinct, since 2002, no one has seen a Yangtze River Dolphin.

Figure 6.2: Baiji White Dolphin.

Source: https://onekindplanet.org/top-10/top-10-worlds-extinct-animals/

Pyrenean Ibex

One of four subspecies of the Spanish Ibex or Iberian Goat that was found in the Iberian Peninsula. The average height of Ibex is 60–76 cm at the shoulder and weight 24–80 kg and they are mainly dependent on grasses and herbs to fulfill their food needs. Historically, they were thought to be 50000 in number, but later in the 1900s, their number was fallen down to even less than 100. Scientists believe factors including poaching and the inability to compete with other mammals for food and habitat leads to the extinction of these species, but the exact cause of the Pyrenean Ibex's extinction is still unknown. In 2000, in northern Spain, the last Pyrenean Ibex was killed by a falling tree.

Figure 6.3: Pyrenean ibex.

Source: https://onekindplanet.org/top-10/top-10-worlds-extinct-animals/

Passenger Pigeon

Since the early 20th century, native to North America, the Passenger or Wild Pigeon has been extinct. It is projected that around 3–5 billion Passenger Pigeons populated the United States when Europeans arrived in North America, but their settlement led to mass deforestation resulting in habitat loss and a reduction in the bird population. The 19th century pigeon commercialized meat as cheap food for the poor, which resulted in hunting on a massive scale. The Passenger Pigeon died out in the wild by around 1900, with the last known individual dying in captivity in 1914.

Figure 6.4: Passenger Pigeon.

Source: https://onekindplanet.org/top-10/top-10-worlds-extinct-animals/

Tasmanian Tiger

The Tasmanian Tiger was a large carnivorous marsupial, which was native to Australia, Tasmania, and New Guinea. Though they look like dogs, the dark stripes over their body gave them the appearance of a tiger, but, in general, they are not related to tigers. In between 1910 and 1920s, the last wild Tasmanian Tiger was killed, with the last captive one dying in Hobart Zoo, in 1936 in Tasmania.

Figure 6.5: Tasmanian Tiger.

Source: https://onekindplanet.org/top-10/top-10-worlds-extinct-animals/

Steller's Sea Cow

In 1741, a naturalist named George Stellar discovered this sea cow on whose name it is now popularly known, Steller's Sea Cow was a large herbivorous mammal. It is believed that Steller's Sea Cow whose weight is around 8–10 tons and which grew to at least 8–9 m inhabited the Near Islands, southwest of Alaska and the Commander Islands in the Bering Sea. Steller's sea cow is a domestic mammal and spent most of its time eating kelp; since they were unable to submerge their enormous body, due to which they were highly susceptible to human hunters. Steller's Sea Cow was hunted to extinction, within 27 years of discovery by Europeans.

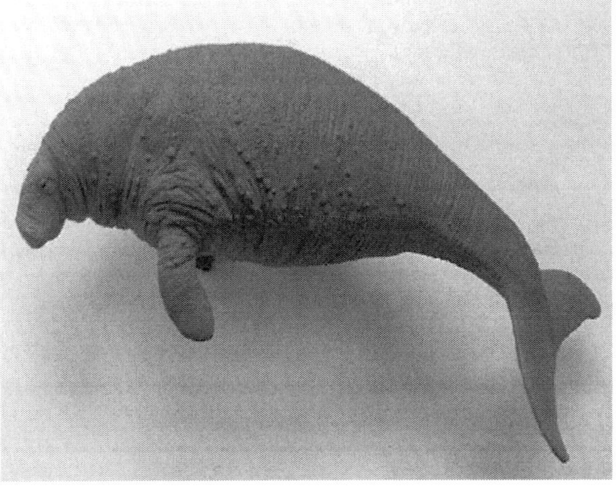

Figure 6.6: Steller's sea cow.

Source: https://onekindplanet.org/top-10/top-10-worlds-extinct-animals/

Great Auk

The great auk is a large and flightless bird and are found in the North Atlantic and as far south as Northern Spain. The average height and weight of great auk is approximately 75–85 cm and 5 kg, respectively. The Great Auk was a powerful swimmer, who helped it to hunt underwater for food. The last colony of Auks was found on the islands of Eldey, and by 1835, they were all killed. The last of these birds were killed by three men who caught it on St Kilda, Scotland, in 1844.

Figure 6.7: Great auk.

Source: https://onekindplanet.org/top-10/top-10-worlds-extinct-animals/

Dodo

Dodo was found in Mauritius. Dodo was a flightless bird, which was about 1 m tall, and their weight is around 10–18 kg. The presence of dodo is only known through the varied illustrations and written accounts from the 17th century; therefore, its exact presence is still a matter of mystery. Due to the availability of abundant food sources (seeds, roots and fallen fruits) and a relative absence of predators, the bird became flightless, as presumed by scientists. In 1598, Dutch sailors first recorded a mention of the dodo. Sailors and their domesticated animals, as well as invasive species, hunted the bird to extinction.

Figure 6.8: Dodo.

Source: https://onekindplanet.org/top-10/top-10-worlds-extinct-animals/

Woolly Mammoth

The woolly mammoth is giant in size. They are believed to be closely related to the present-day elephants. The ancestors of woolly mammoth migrated out of Africa about 3.5 million years ago, spreading across North America and northern Eurasia. The woolly mammoth was around 6 tons in weight and 4 m in height. They were covered in fur, and their curved tusks could easily be up to 5 m long. Through a combination of hunting by humans and the disappearance of its habitat through climate change, the Woolly Mammoth eventually disappeared 10,000 years ago. around 1700 BC, the last of the isolated woolly mammoth populations is believed to have vanished from Wrangel Island in the Arctic Ocean.

Figure 6.9: Woolly mammoth.

Source: https://onekindplanet.org/top-10/top-10-worlds-extinct-animals/

Sabre-toothed Cat

Sabre-toothed cat existed 55 million to 11,700 years ago. They are often called by the name Sabre-toothed Tigers or Sabre-toothed Lions. Sabre-tooth Cats were carnivores named for the extended bladelike canine teeth, which in some species were up to 50 cm long. They were believed to be excellent hunters and hunted animals such as sloths and mammoths, because of their quite bear-like build. These cats could open their jaws at an angle of 120°—almost twice as wide as a modern lion! It is believed the Sabre-tooth Cat's extinction may be linked to the decline and extinction of the large herbivores they hunted, ultimately lead to the extinction of Sabre-toothed cats as they had nothing to depend upon for their feed. Other reasons include climate change and competition with humans.

Figure 6.10: Sabre-toothed cats.

Source: https://onekindplanet.org/top-10/top-10-worlds-extinct-animals/

REFERENCES

1. Biologicaldiversity.org. (2017). *The extinction crisis*. Retrieved October 14, 2017, from http://www.biologicaldiversity.org/programs/biodiversity/elements_of_biodiversity/extinction_crisis/
2. Briggs, J. (1987). *Bio-geography and plate tectonics*. Retrieved October 14, 2017, from http://file:///C:/Users/Cognitive/Downloads/J.C.%20Briggs%20Eds.%20Biogeography%20and%20Plate%20Tectonics.pdf
3. Briggs, J. (1995). *Global biogeography*. Retrieved October 14, 2017, from http://file:///C:/Users/Cognitive/Downloads/John%20C.%20Briggs%20Eds.%20Global%20Biogeography.pdf
4. OneKindPlanet. (2017). *10 animals that are now extinct*|OneKindPlanet. Retrieved October 14, 2017, from https://onekindplanet.org/top-10/top-10-worlds-extinct-animals/
5. ThoughtCo. (2017). *Biogeography provides evidence of evolution*. Retrieved October 14, 2017, from https://www.thoughtco.com/how-biogeography-supports-evolution-249883
6. Wandrei, K. (2017). *Examples of biogeographical proof for evolution*. Retrieved October 14, 2017, from https://sciencing.com/examples-biogeographical-proof-evolution-12944.html
7. Whittaker, R, Araújo, M., Jepson, P., Ladle, R., Watson, J., & Willis, K. (2010). Conservation biogeography: Assessment and prospect. *Diversity and distributions*. Retrieved November 3, 2017, from http://dx.doi.org/10.1111/j.1366-9516.2005.00143.x. doi: 10.1111/j.1472-4642.2010.00660.x
8. Young, K. (2015). Conservation biogeography – Geography – Oxford bibliographies – obo. Oxfordbibliographies.com. Retrieved October 14, 2017, from http://www.oxfordbibliographies.com/view/document/obo-9780199874002/obo-9780199874002-0043.xml

7
CHAPTER

THE CONVENTION ON BIOLOGICAL DIVERSITY (CBD)

"The variability among living organisms from all sources including, inter alia, terrestrial, marine and other aquatic ecosystems and the ecological complexes of which they are a part; this includes diversity within species, between species, and of ecosystems."

"At least 40 per cent of the world's economy and 80 per cent of the needs of the poor are derived from biological resources. In addition, the richer the diversity of life, the greater the opportunity for medical discoveries, economic development, and adaptive responses to such new challenges as climate change. The variety of life is our insurance policy. Our own life and livelihood depends on it."

—The Convention on Biological Diversity

CONTENTS

7.1 INTRODUCTION

The importance of biodiversity was recognized approximately 20 decades ago because of a very important inter-governmental agreement, which was established among several countries. This inter-governmental agreement was named as "Convention on Biological Diversity," which was agreed by 193 countries to support the protection of biodiversity and use natural resources sustainably with equitable profit sharing. Hence, it is widely recognized that global biodiversity is altering at an unprecedented rate because of increasing human-induced changes in the global environment setting. The extent of this change is strongly associated with the processes of ecosystem and the use of natural resources by humans that change in biodiversity is considered as a significant global change. To finish, it could that there are several national and international conventions that are established to minimize the loss of biodiversity, but the changing scenario of climate and increasing human interference in natural cycle is affecting the future state of global environment.

In 1992, the UN Convention on Biological Diversity (CBD) proposed the following three ultimate objectives:

- The conservation of biological diversity
- The sustainable use of its components
- The fair and equitable sharing of the benefits arising out of the use of genetic resources

In 2002, parties of the CBD adopted a strategic plan "to achieve by 2010 a significant reduction of the current rate of biodiversity loss at the global, regional and national level as a contribution to poverty alleviation and to the benefit of all life on Earth." This target was later authorized by the "World Summit on Sustainable Development" and incorporated as a target under the "Millennium Development Goals." It is broadly recognized, however, that the 2010 biodiversity target was not met. The United Nations General Assembly declared 2011–20 the United Nations Decade on Biodiversity, recognizing the importance of this global environment problem.

7.2 HISTORICAL DEVELOPMENT

At global level, there has been a development of various policies for the sustainable development of natural resources since a very long period of time. This is because they prefer avoiding the long-term damage to resources and services offered by the ecosystem. Aspects of biodiversity management have been on the international agenda for many years, although early international environmental treaties were primarily concerned with controlling the excess exploitation of particular species.

In 1972, the United Nations Conference on Human Environment was held in Stockholm, where modern attempts to manage global diversity were first considered. It stated that the conservation of biodiversity is the priority of the convention, and hence, it is necessary to implement rigorous rules and regulations for the safeguard of biodiversity. The Action Plan in Programme Development and Priorities adopted in 1973 at the first session of the Governing Council of UNEP identified the "conservation of nature, wildlife and genetic resources" as a priority area. Furthermore, the international importance of conservation was confirmed by the adoption, in the same decade, of the Convention on Wetlands (1971), the World Heritage Convention (1972), the Convention on International Trade in Endangered Species (1973), and the Convention on Migratory Species (1979), as well as various regional conventions.

To save the environment and prevent biodiversity loss, it is important to put a control over the reasons that results in biodiversity loss. Biodiversity loss mainly occurs due to human-induced actions that cause threat to biodiversity and ecological system. Thus, governmental policies and acts should be enforced to save the environment. However, there are several acts that exist already for the protection of environment: Environment Protection Act 1986, Forest Act 1927, Biodiversity Act 2002, and Wildlife Protection Act 1972. Although all these acts and legislations were established with focus on the conservation of the environment, none of them were focused on the dimensions of biodiversity and ecology preservation.

7.2.1 Issues taken care by the Convention

The convention on biodiversity was framed to deal with various issues related to biodiversity. Some of the issues, which are taken care by this convention, are discussed below:

- This convention measures the inducements related to the conservation and sustainability of biodiversity. It also deals with ecological services provided by biodiversity and its components, so that they could be used in a sustainable manner.
- It also regulates access to all the resources related to genes and traditional knowledge. But for that, prior informed consent of the party is required for providing resources.

- It shares, in a fair and equitable way, the results of research and development and the benefits arising from the commercial and other use of genetic resources with the Contracting Party providing such resources (governments and/or local communities that provided the traditional knowledge or biodiversity resources utilized).
- It provides access to technology. This also includes biotechnology, which could be used and could help governments or local people in providing traditional knowledge on the resources available through biodiversity.
- Technical and scientific cooperation is another main agenda of the convention on biodiversity.
- The coordination of a global directory of taxonomic expertise (Global Taxonomy Initiative) is another major issue tackled by this convention.
- The convention on biodiversity creates awareness among various classes of people, and it also educates people to use resources effectively and efficiently.
- There is adequate Provision of financial resources that helps in funding of this program.
- There is a nationwide reporting on efforts to implement treaty commitments.

7.2.2 National Biodiversity Strategies and Action Plans (NBSAPs)

The Convention on Biological Diversity calls for each participant of the convention to develop a National Biodiversity Strategy and Action Plan (NBSAP). The NBSAP is formulated to take care of all the objectives of the convention at all the levels and in all the sectors of the society. This has been stated in Article 6 of the Convention. Moreover, the national biodiversity strategy throws light on the vision of a country regarding the conservation of biodiversity, and it also focuses on the broad policy and institutional measures that the country should take to fulfill the objectives of the Convention. The action plan of the Convention and its subordinate articles comprises a concrete action that is supposed to be taken to achieve the main objectives of the National Biodiversity Strategies and Action Plans.

Furthermore, according to the Convention, "the strategy should include ambitious but realistic and measurable national targets developed in the framework of the Aichi Biodiversity Targets adopted at the tenth meeting of the Conference of the Parties." Each party in respect to the national circumstances and also the capabilities of each party to handle the situation develops the strategies and action plans. While framing the action plans, it is essential that all sectors whose activities impact on biodiversity be brought into the NBSAP process at an early stage. This is known as mainstreaming, wherein all the groups of stakeholders within biodiversity work together in order to form and implement the NBSAP.

Mainstreaming is also referred as the integration of biodiversity taking into consideration various plans, legislations, programs, policies, and so forth, these plans and policies are essential for the sustainable development and conservation of biodiversity. Examples for various plans and strategies are; National Development Plans; National Strategies for Sustainable Development; Poverty Reduction Strategy Papers; Strategies to achieve the Millennium Development Goals; National Programs to Combat Desertification; National Climate Change Adaptation or Mitigation Strategies; and relevant private-sector policies.

While the NBSAP can take the form of a single biodiversity-planning document, it can also be conceived as comprising a "basket" of elements on. The examples for such documents and baskets are laws and administrative procedures; scientific research agendas, programs, and projects; communication, education, and public awareness activities; forums for inter-ministerial and multi-stakeholder dialogue. The NBSAP is a process that is responsible for increasing the knowledge and information of the society through the evaluation of each phase of implementation made by the Convention on Biodiversity. Finally, an ongoing review on these action plans and strategies is also important to provide, so that time-to-time improvements are made (if required).

7.2.3 International Day for Biological Diversity: May 22

The United Nations proclaimed May 22 as the International Day for Biological Diversity (IDB). This day was kept especially for biological diversity to increase the understanding and awareness of biodiversity issues among all the generations of the society. On this day, governments and citizens from around the world take time to discover the biodiversity that surrounds them, and they celebrate some of the successful activities, which help in supporting biodiversity conservation and its sustainable use.

International Day for Biological Diversity is an important part of the Program of Work on Communication, Education and Public Awareness (CEPA). The Secretariat of the Convention on Biological Diversity (CBD) organizes a big program on this day every year with a different theme, including kids, adults, and other party members of the Convention. To help highlight the key issues pertinent to the given theme, the Secretariat forwards the agenda to all the parties and other relevant partners. This agenda is basically circulated to create awareness on some crucial issues, on the basis of the resources available to them. Past, present, and future International Days for Biological Diversity are:

2012: Marine and Coastal Biodiversity

2011: Forest Biodiversity

2010: Biodiversity, Development and Poverty Alleviation

2009: Invasive Alien Species

2008: Biodiversity and Agriculture

2007: Biodiversity and Climate Change

2006: Protect Biodiversity in Drylands

2005: Biodiversity: Life Insurance for our Changing World

2004: Biodiversity: Food, Water and Health for All

2003: Biodiversity and Poverty Alleviation and Challenges for Sustainable Development

7.3 BIODIVERSITY HOTSPOTS ACROSS THE WORLD

A need to identify the most important areas in the world to have global actions–led biodiversity conservation managers and scientists to propose what is called as "Biodiversity Hotspots": biodiversity hotspot is a biogeographic region that is both a reservoir of biodiversity and is threatened with destruction. According to Myers (1988), there are 10 tropical forests that are considered as major hotspots in the world. These hotspots are evaluated on the basis of surprising plant endemism high level of habitat loss. However, few numbers of hotspots were added to list in the later stage. To qualify as a hotspot, a region must contain at least 1,500 endemic vascular plants and have 30% of its original vegetation. It is thus revealed that at present, approximately 34 biodiversity hotspots are discovered. Finally, there is a periodical update in the hotspots, which provides a clear picture of natural wealth present on the planet in these significant areas. The richness of hotspots can greatly influence the biodiversity loss.

The regions that qualify for hotspot criteria hold at least 1,500 endemic species of plants, but in the present scenario, each hotspot has lost 70% of its original habitat. If the past data are considered, the total hotspots covered approximately 23.7 million square kilometers of land that accounts for almost 15.9% of the earth's surface. However, because of extreme habitat destruction in this area, the hotspot now contains only 3.4 million square kilometers of the total land in the world, which accounts for 2.3% of the earth's surface. It is revealed that more than 85% of the habitat present in hotspots has been destroyed at present.

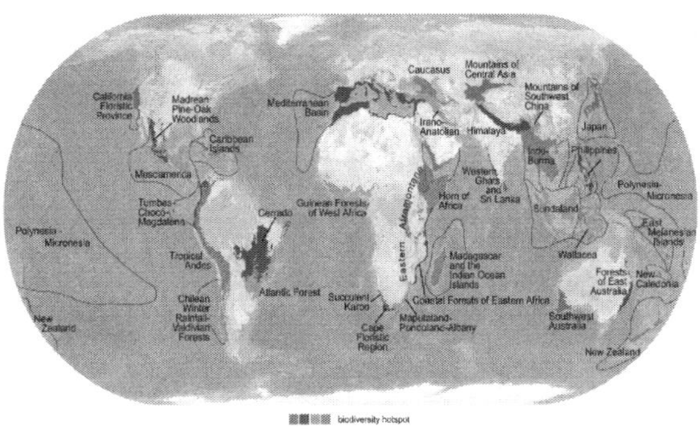

Figure 7.1: The Biodiversity Hotspots.

(Source: Conservation International Foundation www.conservation.org)

Hotspots around the world are estimated to hold approximately 152,000 species of plants and 22,939 terrestrial vertebrates including mammals, birds, reptiles, and amphibians. If the individual vertebrate class is calculated in the hotspot region, it accounts for 46% of the reptiles, 35% of the birds, 35% of the mammals, and 59% of the amphibian species in total. If threatened species are considered in hotspot regions, the hotspot region contains 79% of the threatened amphibians, 60% of mammals, and 63% of the bird species are threatened in the region.

Table 7.1 shows the location of the major biodiversity hotspots according to high biodiversity and high endemism values for conservation and for both plants and animal groups, and Table 7.2 shows the number of families and genera in the hotspots with the largest amount of endemic species of vertebrates.

Table 7.1: Plant and Vertebrates Species in Biodiversity Hotspots

O: stands for Occurring in Biodiversity Hotspot

E: stands for Endemic to Biodiversity Hotspot

Source: https://www.researchgate.net/publication/225989074_Global_Biodiversity_Conservation_The_Critical_Role_of_Hotspots#pf13

Hotspots	**Plants**		**Birds**		**Reptiles**		**Freshwater Fishes**	
	O	E	O	E	O	E	O	E
Tropical Andes	30,000	15,000	1,728	584	610	275	380	131
Tumbes-Choco-Magdalena	11,000	2,750	892	112	325	98	251	115
Atlantic Forest	20,000	8,000	936	148	306	94	350	133
Cerrado	10,000	4,400	605	16	225	33	800	200
Chilean Winter Rainfall and Valdiv	3,892	1,957	226	12	41	27	43	24
Mesoamerica	17,000	2,942	1,124	213	686	240	509	340
Madrean pine-Oak Woodlands	5,300	3,975	525	23	384	37	84	18
Caribbean Islands	13,000	6,550	607	167	499	468	161	65
California Floristic Province	3,488	2,124	341	8	69	4	73	15
Guinean Forests of West Africa	9,000	1,800	793	75	206	52	512	143
Cape Floristic Region	9,000	6,210	324	6	100	22	34	14
Succulent Karoo	6,356	2,439	227	1	94	15	28	0
Maputal and-Pondoland–Albany	8,100	1,900	541	0	205	36	73	20
Costal Forest of Eastern Africa	4,000	1,750	636	12	250	54	219	32
Eastern Afromontane	7,598	2,356	1,325	110	347	93	893	617
Horn of Africa	5,000	2,750	704	25	284	93	100	10
Madagascar and the Indian Ocean I	13,000	11,600	313	183	381	367	164	97
Mediterranean Basin	22,500	11,700	497	32	228	77	216	63
Caucasus	6,400	1,600	381	2	87	20	127	12
Irano-Anatolian	6,000	2,500	364	0	116	13	90	30

Mountains of Central Asia	5,500	1,500	493	0	59	1	27	5
Western Ghats and Sri Lanka	5,916	3,049	457	35	265	176	191	139
Himalaya	10,000	3,160	979	15	177	49	269	33
Mountains of Southwest China	12,000	3,500	611	1	94	15	92	23
Indo-Burma	13,500	7,000	1,277	73	518	204	1,262	553
Sundaland	25,000	15,000	771	146	449	244	950	350
Wallacea	10,000	1,500	650	265	222	99	250	50
Philippines	9,253	6,091	535	185	235	160	281	67
Japan	5,600	1,950	368	15	64	28	214	52
Southwest Australia	5,571	2,948	285	10	177	27	20	10
East Melanesian Islands	8,000	3,000	365	154	114	54	52	3
New Zealand	2,300	1,865	198	89	37	37	39	25
New Caledonia	3,270	2,432	105	23	70	62	85	9
Polynesia-Micronesia	5,330	3,074	300	170	61	31	96	20
Forests of East Australia	8,257	2144.0	632	28	321	70	80	10

Table 7.2: Hotspots with the Largest Number of Endemic Higher Vertebrate Taxa (including all mammals, amphibians, birds, freshwater fishes, and reptiles)

Source: https://www.researchgate.net/publication/225989074_Global_Biodiversity_Conservation_The_Critical_Role_of_Hotspots#pf13

Rank	Genera	Families
1	Madagascar and the Indian Ocean Islands (175)	Madagascar and the Indian Ocean Islands (22)
2	Eastern Afromontane (119)	Philippines (16)
3	Tropical Andes (103)	Japan (8)
4	Sundaland (97)	Sundaland (7)
5	Mesoamerica (78)	Caribbean Islands (6), Chilean Winter Rainfall and Valdivian Forests, Wallacea, New Zealand, New Caledonia (4)
6	Indo-Burma (68)	
7	Caribbean Islands (65)	

8	Atlantic Forest (63)	
9	Wallacea (62)	
10	Philippines (45)	Mesoamerica, Indo-Burma, and Polynesia–Micronesia (3)

"Introduced predators have devastated island hotspots, where species evolved in the absence of domestic cats and rats and other invasive predators" as stated by Steadman in 1995. "Introduced plants are having massive impacts on hydrology and biodiversity in some hotspots, particularly those having Mediterranean-type vegetation" as stated by Groves and di Castri in 1991. Misuse for protein (e.g., bushmeat), for medicine, and for the pet trade 1 Global Biodiversity Conservation: "The Critical Role of Hotspots 17 threatens species in all hotspots, particularly the Guinean forests of West Africa (Bakarr et al. 2001), Madagascar, and hotspots in Southeast Asia." However, an overlay of the hotspots with protected areas with defined boundaries from the World Database on Protected Areas (IUCN and WCMC 2009) exposes that 12% of the unique area of the 35 hotspots is under some form of protection, while 6% is classified as IUCN category I–IV protected area. As boundaries for many protected areas have not been systematically compiled, these numbers are underestimated, and they certainly overrate the land area that is succeeded efficiently. Yet the fraction of hotspots protected is less expressive than the locations themselves. To conserve biological hotspots, efforts are taken to ensure the long-term persistence of the areas already protected, as well as strategically add new protected areas in the highest priority unprotected habitats that remain intact as indicated by systematic efforts to identify gaps in networks of protected areas. Another major challenge is maintaining the resilience of hotspots in the face of climate change. However, the distances that species must move are likely to be shorter in mountainous terrain and longer in flatter regions, because of the nature of climatic gradients. Finally, mountains are more expected to have habitat breaks that make species dispersion more difficult.

It is, thus, very necessary to protect those sites where species currently exist, chiefly the "Key Biodiversity Areas" where species are at the greatest risk. As stated by Ricketts et al. in 2005, "The hotspots, in fact, harbor 81% of the global total 595 Alliance for Zero Extinction sites locations harboring the sole remaining populations of the most threatened species." It is, thus, very important to protect the existing species now because if we cannot even save these species then we will not be granted with any other chance later. However, this is only the beginning. We must also protect habitats where species will be in the future, as well as provide "stepping stones" to enable movement to these new series. Some of those stepping zones are:

- Relate biodiversity to everyone's daily needs so that the ownership for saving biodiversity is spread and everyone feels responsible for stopping loss.
- Ensure science, policy, and politics play an equal role in decision-making.
- Economics make a great case to argue for conservation action. But such arguments should be translated into national and local actions to realize the potential of biodiversity.
- Revalue our choices and life-styles to provide space for the species that inhabit this Earth.

7.4 THE CBD AND OTHER INTERNATIONAL AGREEMENTS

Given the need to establish synergies with other international bodies that also aim to preserve the biological diversity or part of it, the CBD has established important global agreements with Migratory birds, CITES (International Trade of Endangered Species), Ramsar (Wetlands) and UNCCD (climate change), among others of global importance:

7.4.1 CBD and Migratory Birds

The problematics around the migratory birds led the CBD to establish a specific agreement. It was informed from Western Port and many other ports that are home to many migratory shorebirds, including the red-necked stint, the eastern curlew, the curlew sandpiper, the great knot, and the red knot. Many of these bird species breed in the Arctic during the northern summer then migrate up to 12 000 kilometers to spend the annual non-breeding season in southern Australia's intertidal wetlands. It also occurs in North America and South America. In Australia, it was informed that over 30 years' worth of data collected regularly by volunteers at sites around Western Port suggest a concerning trend, with nearly all migratory shorebird species declining in numbers. Species that have seen particularly dramatic declines include the eastern curlew, the curlew sandpiper and the red knot.

On the other hand, the major route taken by these birds during their migration from the Arctic to southern Australia is known as the East Asian–

Australasian Flyway. The CBD was advised that the absence of changes to Australian shorebird sites in recent years suggested that the species declines were attributable to changes outside Australia. Moreover, the CBD was informed that the loss of habitat in East and South East Asia is suspected to be the primary cause of declines in Australia's migratory shorebird numbers. Tidal mudflats around the Yellow Sea are key stop-over sites for many birds on the East Asian-Australasian Flyway, and access to adequate food sources in these areas is essential for birds to continue traveling the long distance to their Arctic breeding grounds. Finally, the CBD heard that between 50 and 60% of the tidal zone around the Yellow Sea has been reclaimed, and BirdLife Australia provided several examples of recent large-scale industrial developments on sites, which were once important feeding grounds.

It was suggested that, although Australia has bilateral agreements in place with the governments of Japan, China and the Republic of Korea for the protection of migratory birds, the agreements have not been entirely effective in preventing the destruction of key stop-over habitats for migratory birds. Given that climate change is expected to provide an even more challenging set of circumstances for bird species, the Committee considers that there would be benefits in mitigating other pressures on migratory birds wherever possible. The Committee therefore notes concerns about the adequacy of international agreements for the protection of migratory bird habitats and will consider this matter carefully in the context of its inquiry.

7.4.2 CBD and the Ramsar Protection of Wetlands

RAMSAR is the International Convention to protect wetlands of global value. Given the high importance of wetlands as regulators of climate change, as regulators of floods, biogeochemical changes, among many others, the CBD analyzed how Ramsar could provide support to the development of data needed for the CBD. RAMSAR acts as the lead partner implementing the CBD in the Strategic Plan 2011–2020, which has placed a great value on water and water-related ecosystem services.

7.4.3 CBD and CITES

CITES is the Convention of International Trade in Endangered Species of Wild Fauna and Flora; therefore, the cooperation between CBD and CITES took place in September 2016, although this cooperation started on April 2004.

CONCLUDING REMARKS

Having received evidence from other museums at public hearings held earlier in the inquiry process, the Committee was pleased to have the opportunity to visit a natural history museum in Melbourne. Highlights of the visit included seeing firsthand the scope of the Melbourne Museum's collections and research facilities, and hearing about how technological advances are changing the way in which the museum conducts its research and engages with visitors and the general public. Moreover, noting the importance of birds as key indicators of environmental health, the Committee appreciated the opportunity to inspect key shorebird habitat in Western Port. The site inspection highlighted the vulnerability of birds to existing threats and the anticipated effects of climate change. To end, the Committee wishes to thank Melbourne Museum and BirdLife Australia, whose representatives provided invaluable assistance and information throughout the Committee's site inspections in Victoria.

REFERENCES

1. CBD – Convention on Biological Diversity. (n.d.). Retrieved October 14, 2017 from https://www.cbd.int/undb/media/factsheets/undb-factsheets-en-web.pdf
2. CBD.int. (2017). *Status and trends of global biodiversity*. Retrieved October 14, 2017 from https://www.cbd.int/gbo1/chap-01.shtml
3. Mittermeier, A. R. (2011). *Global biodiversity conservation: The critical role of hotspots*. Retrieved October 14, 2017 from https://www.researchgate.net/publication/225989074_Global_Biodiversity_Conservation_The_Critical_Role_of_Hotspots#pf13
4. UNDP. (n.d.). *International day for biological diversity 22 May*. Retrieved October 14, 2017 from http://www.un.org/en/events/biodiversityday/convention.shtml

8
CHAPTER

ANALYTICAL APPROACHES, ADVANCES, AND FUTURE DIRECTIONS

CONTENTS

ABSTRACT

Biogeography is, as of now, creating space. However, an established inception does not really guarantee a cutting-edge personality in the proceeding with battle struggle between the "hard" and "delicate" sciences. The diverse structures and rates of advancement in the land and organic sciences have given biogeography a sort of twofold character, which has served just to confound and postpone its foundation as a different science. The chapter talks about an extensive variety of themes, for example, biogeography and the biological systems, bioclimates, and vegetation examination, pinpointing their present phases of improvement and future integrative part in the arrangement of both pure and connected social and natural issues.

It, likewise, exhibits a set of cases from many fluctuated spatial procedures and analytical methodologies being utilized by geographers, environmentalists, and biogeographers to study plant and creature appropriations, to survey forms influencing the observed patterns at chosen spatial and temporal scales, and to talk about these cases with a solid applied spatial and temporal structures.

8.1 INTRODUCTION

Nowadays, human activities progressively influence the earthly biosphere, bringing about natural surroundings misfortune and corruption, which, at last, hinder environment capacity and biological system administrations. Because of the velocity of this procedure, there is an extreme need to identify and foresee changes in the common habitat and survey the spatial appropriation of important locales and territories.

On the other hand, European and national environmental regulations frequently require appraisals of the environmental impacts of land utilization planning. Not just point-by-point information of the current examples of species lavishness and circulations (in the future biodiversity), yet, in addition, precise spatial forecasts for all the more ineffectively known districts are required. The test in protection and land utilization planning is that biodiversity information, as exact as could be expected under the circumstances, is required inside a constrained measure of time and over significantly expansive regions. Tragically, such information is frequently meager and costly to secure by customary field inventories.

Along with biogeographers and landscape ecologists, ecological analysts, organizers, and policy-makers are swinging to the discipline of remote sensing (RS) and geographic information (GI) science. RS–GI science is thought to be fit for giving the required systems and information for creating savvy evaluations, models, and projections of biodiversity reactions to natural change over expansive regions. Desires on RS–GI systems have been supported by the developing choice of remotely detected information (regularly openly available) sources. Right now, however, information chronicles give expanding amounts of multi-transient and multi-ghastly remotely detected symbolism.

On the other hand, remote detecting creates a momentous cluster of biogeographically significant estimations and, in addition, the ability to identify normal and human-prompted arrive cover changes. Utilized as a part of an integrative mode, RS–GI information can give data about both verifiable and current land cover factors influencing biodiversity designs. Connections between species disseminations and remotely detected information, especially in conjunction with topographic, soil and bedrock data, can be utilized to foresee the circulation of species over expansive ranges. However, remotely detected imagery depicts land-cover changes at spatial scales from neighborhood to mainland, enabling one to screen the pace of natural surroundings transformation. These estimations of living space misfortune can be utilized as a part of giving quantitative appraisals of biodiversity misfortune using species environment demonstrating. The test is to go above and beyond and give a more point-by-point comprehension of which species are being lost and why.

So how might we best match existing and rising RS and GI advancements to parameters that have clear ramifications for creatures and biological communities?

Up to this point, the advantages of remote detecting have not been completely used in biodiversity readings. In addition, there are a few basic issues in applying RS-GI based biodiversity models that have been deficiently investigated. These incorporate the pertinence of various RS informational indexes in biodiversity evaluations, and the impact of scale, measurable systems, and model unpredictability on RS–GI-based biodiversity displaying. In this chapter, novel methodologies and procedures to coordinate RS and GI into biogeographical research is displayed. The issues talked about in this chapter can have significance in a few fields of use of spatial information in physical topography and related ecological sciences.

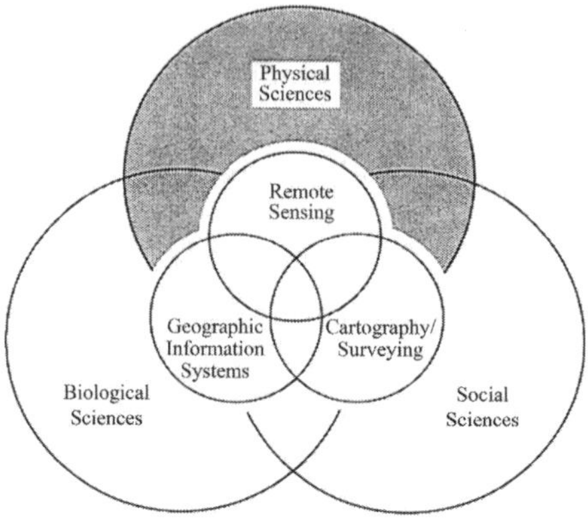

Figure 8.1: Venn diagram of physical, biological, and social sciences with remote sensing, GIS, and cartography? repeat, bad image?

Source: Jensen, 2000

8.2 AN IMPORTANT ROLE OF BIOINFORMATICS AND DATA SCIENCE ON BIODIVERSITY

As new advances develop and our capacity to address the dynamic idea of biodiversity progresses, so too does the accessibility of gigantic measures of information and progressively modern tools that can investigate the information, which thus opens up a universe of chance for understanding biogeographic examples. Data on ecological factors are presently remarkably complex, with worldwide data on geology, hydrography, soils, bioclimate, and so forth. Geographic data on biodiversity, in general, have slacked, with the comprehension of even the best-known species being substantially coarser than other ecological data, named as "Wallacean shortfall."

In any case, impressive progress has been made of late. Specifically, one push to address this deficit is the *Map of Life,* which plans to give fine-scale species event data by incorporating crosswise various information sorts as accessible, including local agendas, master extend maps, displayed dispersions, and phylogenetic data. Thus, major local endeavors to coordinate topographically important natural information the *Atlas of Living Australia*, which gives spatial biodiversity data together with inventive tools such as REBIOMA for Madagascar.

The measure of uninhibitedly accessible information, for example, around the world, particularly museum-based records, is expanding quickly.

Real-example information aggregators incorporate the *Global Biodiversity Information Facility*, the *Paleobiology Database*, and *Ocean Biogeographic Information System*. Regional aggregators incorporate *Canadensys*, the *US Virtual Herbarium*, and *New Zealand Virtual Herbarium*. Moreover, there are expanding quantities of event databases such as *MaNIS*, *FishNet*, *HerpNet*, and *ORNIS*.

Striking species record and information databases are *AmphibiaWeb*, *FishBase*, and *Neotoma*. Extra depositories incorporate *Dryad* for information and *MorphBank* for pictures. A considerable lot of these have applications or interfaces to permit simple access for people in general (e.g., *AmphibiaWeb*). Moreover, different bioinformatics endeavors went for coordinating and serving information on biodiversity incorporate *Data Observation Network for Earth*, the *Encyclopedia of Life*, *GEO Biodiversity Observations Network*, activities associated with the *Convention on Biodiversity* and the *Intergovernmental Platform on Biodiversity and Ecosystem Services* including the IUCN list of undermined and jeopardized species, and the Global Invasive Species Database. However, notwithstanding data on examples and related natural information, the accessibility of genetic and genomic information has detonated throughout the most recent decade or something like that, with huge measures of information accessible through the *National Center for Biotechnology Information* (NCBI, e.g., Genbank and the Gene Expression Omnibus).

Notwithstanding creating openness of existing information, a few endeavors are in progress to improve information obtaining, including the utilization of national researchers and crowd-sourcing for digitization of specimen labels. Extra citizen science endeavors are centered around common history perceptions (e.g., *iNaturalist* and *eBird*). Unmistakably, the monstrosity of the information that is getting to be plainly accessible holds gigantic guarantee, in spite of the fact that the entanglements of utilizing this information uncritically is as of now very much perceived. Key significance is to ground the information examination and translation inside a strong logical structure.

8.3 REMOTE SENSING AND GIS IN BIOGEOGRAPHY: SPATIAL AND TEMPORAL VIEWPOINTS

Since the 1970s, remote sensing has progressively turned into a noteworthy data source for many ecological controls at a scope of spatial and temporal scales. These incorporate biogeography and biology; however, the inclination gives off an impression of being toward the last mentioned. Remote sensing is significantly more imperative in Biogeography. This can be strengthened by the contention that a few parts of spatial ecology (i.e., habitat fragmentation in landscape ecology), and worldwide perspectives on phonological changes in vegetation could not have advanced to their present levels of refinement without parallel improvements in remote sensing. However, the utilization of remotely detected information in biogeography and nature are broad and are expanding.

There are many factors behind this expansion such as more progress in remote sensing technologies, increasing number of analysts having access to the remotely detected information and new inquiries arriving to the biogeography.

Newton et al. (2009) in an audit of 158 remote sensing-based papers published in 2004–08 in the journal *Landscape Ecology* demonstrated that 46 and 42% depend on Landsat TM and MSS information, respectively. Just 1% utilized IKONOS information, with a large portion of that number utilizing Quickbird information. Moreover, low-spatial resolution data were ineffectively denoted (e.g., AVHRR only accounted for 4% of studies), and information from space offices outside. Europe and North America were additionally inadequately depicted (e.g., IRS-LISS, 1%). Microwave were just utilized as a part of 0.5% of studies. They additionally stretch that a considerable lot of the applications addressed officially existing inquiries in scene biology. In rundown, this recommends it is the accessibility of information and handling programming, and the need to distribute that is prompting expanded applications. On the off chance that this contention remains constant in different territories of biogeography and nature, numerous open doors exist for this and future ages of biogeographers to get a handle on the capability of new sensors and take a gander at applying remotely detected information in novel approaches to answer basic biogeographical and environmental inquiries.

As far back as the beginning of satellite remote sensing in the 1970s, imagery obtained by remote sensors was utilized as a part of biogeographical and environmental examinations. Often, it has given topical data, for example, land cover or gauges of biophysical and biochemical properties of vegetation. Remotely detected information has additionally been utilized to give data on major abiotic factors, for example, fire and wind, which affect biogeographic procedures and examples. However, the utilization of remote sensing is obvious in every single significant biome. Significant consideration has focused on tropical forests, savannas, and prairies and deserts. With tropical backwoods, for instance, there has been extraordinary enthusiasm for activities, for example, the observing of biome-level change, dry spell effects, fragmentation, and biomass.

Current methodologies in using RS in biodiversity demonstrations can be classified in two principle sorts: 1) natural surroundings mapping utilizing RS information, and expectations of species dissemination in light of living space necessities; and 2) coordinated connections between ghastly brilliance esteems recorded from remote sensors and species distributions recorded in field studies. These methodologies require the utilization of data gathered from the field, and also from the remote sensor. It is fundamental however, to build up the quality of the linkages between these various types of data gathered on various scales. For instance, the level of correspondence between natural surroundings maps and species conveyances relies upon the level of environment outline. Such connections ought to be advanced to get greatest data on biodiversity.

On the other hand, in examination, a noteworthy confinement of ghastly esteems biodiversity models is that the connection between spectral reflectance and species appropriations depends on the species being referred to, but also the area. By and large, little direction exists for agents concerning the appropriateness of these strategies for biodiversity evaluations in various living spaces. Subsequently, there is a most extreme need to assess the appropriateness of these two primary sorts of RS data in demonstrating biodiversity on various spatial scales in various landscape mosaics.

8.3.1 The Effect of Scale on the Biogeographical Models

The issues and effects of scale for quite some time have been a focal point in biogeographical, geomorphological and landscape ecological exploration. The spatial scale on which species dissemination displaying is attempted is of principle significance for the outcomes and surmising's biogeographical examinations. The idea of spatial scale comprises of two vital qualities: the unit of inspecting and the geographical space secured. The primary characteristic is characterized by "grain" (or "determination") and "center," grain being extent of the basic scientific unit and center of the territory represented by each data point. The second characteristic is "extent," describing the land area over which examinations are made.

In particular, the decision of spatial determination in displaying activities can specifically influence the natural understandings resulting from the outcomes and the ease of use of RS-GI based models. On the off chance that the reaction of an offered species to ecological elements shows demonstrates large-scale affectability, relative examinations over a scope of scales can show the resolutions at which the species-condition patterns are most precisely displayed. Be that as it may, we do not have the learning of how the degree of natural surroundings data and determination of information influence the execution of RS-GI construct territory relationship models with respect to various spatial scales.

Analysts have reached distinctive decisions about the handiness of territory relationship models for anticipating species nearness and non-attendance. This distinction likely comes from the inability to perceive the impacts of spatial scale. With a specific end goal to build the believability of the species conveyance models, the investigation ought to be led on the scale on which the objective species most unequivocally reacts to the ecological variety. ID of the most appropriate determination for species dissemination displaying regularly requires considers directed over a scope of spatial scales. Krawchuk and Taylor (2003) contended that reviews looking at species reaction to the earth performed at a solitary scale may bring about constrained conclusions and possibly hurtful administration suggestions.

8.3.2 The Effect of Modeling Techniques and Complexity on RS-GI based Models

Off late, species appropriation demonstrating has turned out to be one of the key issues biogeography, preservation planning as well as in landscape ecology. This advancement depends on two patterns: development in the accessibility of remotely sensed (RS) information and improvement of GI methods, and the improvement of novel measurable strategies that can be connected to an extensive variety of biological frameworks.

On the other hand, RS-GI based biodiversity displaying readings have as a rule been directed by utilizing just a single demonstrating approach. Apparent variation is normal from utilizing distinctive systems in light of the fact that diverse models utilize an assortment of presumptions, calculations and parameterizations. Subsequently, when contemplated utilize a solitary demonstrating strategy there is no data of whether the chosen technique gives the best prescient precision to the specific informational collection utilized. Just a couple of endeavors have been made to assess the capacity of various strategies to recognize in complex RS–GI numerous relapse settings including numerous possibly imperative informative. Thusly, deficient direction exists for agents concerning the appropriateness of various demonstrating approaches for species circulation displaying and other biodiversity evaluations.

Presently, there are few diverse demonstrating approaches accessible for biogeographical displaying, going from generalized linear models (GLMs) and generalized additive models (GAMs), to manage based procedures, and novel relapse systems, for instance multivariate adaptive regression splines (MARS). GLMs are scientific augmentations of linear models which can deal with non-linear connections and distinctive sorts of statistical error disseminations such as Gaussian, Poisson, Binomial, or Gamma. GAMs are adaptable information-driven nonparametric augmentations of generalized linear models that permit both direct and complex added substance reaction bends to be fitted.

Classification tree analysis is a govern-based technique that has once in a while been connected in RS-GI based demonstrating studies. CTA does not depend on from the earlier speculations about the connections amongst indicator and reaction factors. The strategy creates a binary tree through double recursive apportioning, a procedure that parts a hub in view of genuine/false answers about the estimation coming about subsets. Simulated neural systems are intricate non-parametric strategies that have been appeared to give exceptionally adaptable capacity approximated to any information. Neural systems can give way to fabricate exact models when the utilitarian type of the hidden conditions is obscure.

Numerous versatile relapse splines are a generally new procedure that consolidates traditional direct relapse, numerical development of splines and double recursive parceling to create a nearby model in which connections amongst reactions and indicators are either straight or non-straight. MARS has been upheld as a powerful strategy to show complex spatial information that can incorporate a moderately high number of factors. The prior outcomes and in addition the investigations by Moisen and Frescino (2002), Thuiller (2003) and Segurado and Araújo (2004) recommend that the utilization of novel non-parametric procedures frequently brings about a marginally preferable model execution over the parametric methodologies. Non-parametric procedures have been reprimanded in light of the fact that they can create an intricate arrangement of capacities, while comparing parametric capacity may catch the vast majority of a similar variety.

Remote sensing and GI techniques give a surmised cost-productive strategy to evaluate the biodiversity status of wide regions on a wide scale. RS–

GI-based modeling approaches deserve further consideration biogeography and connected land utilize planning. Be that as it may, there are additionally numerous potential weaknesses in utilizing satellite imagery and different sorts of spatial information as a surrogate for species lavishness—how to suitably gather source information for demonstrating, specialized issues in handling satellite imagery, methodological traps in relapse models and so on. The mix of biogeographical displaying, RS and GIS innovation requires regularly transdisciplinary aptitudes between topography, environment, statistics and geo informatics. In this manner, the traps for the abuse and manhandle of the GIS innovation with high count limit are extremely self-evident. We ought to be very mindful and not and not make a difference the consequences of prescient models and extrapolations too uncritically, and not without a decent geological information of the examination range. In entirety, wide-based comprehension of geoinformatics and related issues is required by spatial information makers and modelers.

8.3.3 Downscaling Species Atlas Distributions

One of the impediments in utilizing biogeographical informational collections, commonly species' distribution atlases, in local planning is their coarse determination (regularly from 10 to 50 km) with respect to the necessities of perception and land use planning. Because of this, endeavors have been made to downscale species distributions to better resolutions. The term "downscaling" portrays a technique in which data about a procedure with a specific trademark scale are derived from different procedures with bigger scales.

However, general confinements, hypothesis and routine with regards to downscaling techniques are very much depicted in the writing. In any case, just minimal observational work has been led in geology to utilize coarse-grained information so as to display biogeographical procedures at better spatial determination, however observe special cases in Araújo & al. (2005) for species circulation demonstrating in the United Kingdom. Deliberate examinations utilizing downscaling strategies for a few diverse biogeographical procedures using free model alignment and model test zones are to a great extent lacking. All things considered, the possibility of measurable downscaling of the conveyance of vegetation has been deficiently examined.

Downscaling has been finished by joining remotely detected information and master sentiment to allot species to natural surroundings or land-cover classes considered reasonable for species. One of the issues with this approach is that there might be inadequate master learning of species-living space connections for some species and zones. An option is to utilize exact demonstrating systems that investigate the connection between species dissemination and sets of indicator factors to downscale dispersions of species to better spatial determination. One significant presumption in this approach is that the principle drivers of species dissemination at coarse determination are additionally vital determinants of dispersions at better resolutions.

Finally, late investigations have shown huge fluctuation in downscaled demonstrate forecasts. These outcomes feature the requirement for additionally research to test and enhance the approach in order to expand trust in display forecasts. Measurable downscaling has vital confinements, and take-offs from display suspicions are probably going to fluctuate for various species and districts. Specifically, it is critical to examine which species are ineffectively displayed, where and why. Without such an appraisal, it is troublesome completely to survey the ease of use and exactness of downscaled models in true circumstances.

8.4 THE EMERGENCE AND PROMISE OF FUNCTIONAL BIOGEOGRAPHY

Biogeography is the investigation of the distribution of species and biological communities crosswise over space and time and of the fundamental biotic and abiotic variables, systems, and procedures. It is integral to blending little and huge scale examples of species' reactions to worldwide natural change and giving a window to evaluate the significance of earth and developmental history, and in addition changing biotic and abiotic causes that motivate the present circulation of taxa. Generally, biogeography has been established in the species idea. Moreover, researchers long back constructed worldwide maps of species' reaches and species decent variety patterns, which enabled them to create, test, and approve unmistakable natural hypotheses at the foundation of biogeography. To end, biogeography, in any case, has generally grown independently from biological community nature and earth framework science, yet we contend that one of the colossal difficulties of 21st-century biogeography is to give hypothetical baselines and devices for the comprehension and forecast of environment reactions to ecological changes regarding species structure and biogeochemical cycles (water, carbon, supplements and vitality).

During the past few years, new directions in biogeography have been supported by measurable and computational advances, and new species appropriation models have given maps of need zones to preservation and projections of the impacts of worldwide change on species variety range for the following century. In any case, expectations of how environmental groups react to past and anticipated future atmospheres have progressively been tested because of the impediments of species-based methodologies, including reasonable and specialized troubles in fusing species communications, dispersal restrictions, and species' adjustments into prescient models. For sure, on the off chance that we regard species as subjective substances, at that point we have to see every single one. Interestingly, quality-based biogeography can help display species collaboration, dispersal capacity, and physiological resilience all the just and nonexclusively. All the more exactly, on the off chance that we cluster species along the same characteristics tomahawks, at that point knowing cross-species connections, for example, between plant physiological resilience to dry spell and distance across of xylem vessel, can help figure the adjustment in species creation of natural groups in light of variety in abiotic conditions like water accessibility.

Contrary to species-based biogeography, earth science models directly address changes in biogeochemical cycles at a global scale but using a simplistic representation of biodiversity in most cases: few plant functional types (PFT), characterized by mean eco physiological characteristics, are used per biome in models dedicated to simulate and forecast the consequences of global changes on biogeochemical cycling. However, in biodiversity–ecosystem functioning research, the taxonomic composition of an ecosystem is identified as a key driver of ecosystem functioning.

More recently, the functional component of biodiversity, that is, the diversity of forms and functions, has been recognized as the missing link between biodiversity patterns and biogeochemical cycles and perhaps is a core driver of ecosystem services. Given these insights from ecosystem ecology, there is a growing consensus on the need for a better representation of biodiversity—in particular the composition of species, forms, and functions—in earth science models, as stressed in this issue by Reichstein et al.

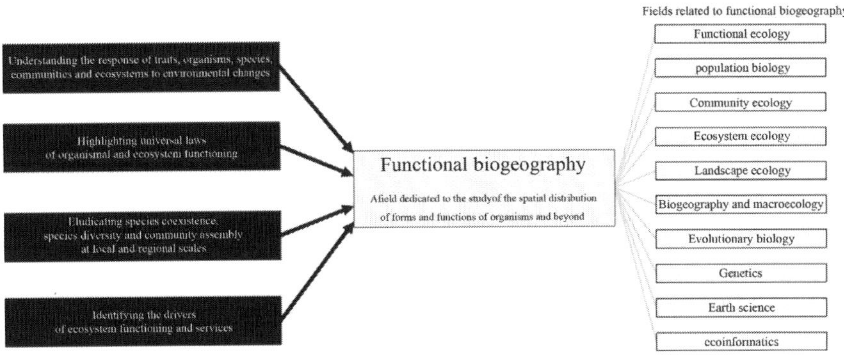

Figure 8.2: The emergence and promise of functional biogeography? repeat bad image?

Source: Proceedings of National Academy of Sciences

8.4.1 Biogeography beyond Species: Functional Traits as a Common Currency across Biological Organizational Levels and Taxonomic Groups

A valuable way to deal with functional biogeography is to utilize a typical currency, specifically utilitarian characteristics. Useful qualities are morphological, physiological, phonological, or behavioral highlights measured on living beings that can at last be connected to their execution. In attribute-based biology, it has been demonstrated that it is conceivable to total practical characteristics measured on life forms to clarify the working of populaces, groups, biological communities, and past. For example, at the group level, the utilization of community-weighed means (CWM, i.e., the normal characteristic estimation of a representing impacts of species plenitude inside groups) is a promising tool to precisely biological community capacities, for example, plant essential efficiency.

In general, utilitarian attributes can be executed into pretty much complex integrative capacities (e.g., CWM is a basic integrative capacity) to scale up from organs to higher hierarchical levels including environments and biomes. Quality based methodologies have additionally been broadly used to depict the decent variety of structures and capacities inside an investigation unit—frequently named practical assorted variety sensu lato—

utilizing distinctive separation measurements (e.g., difference based) and how it scales spatially.

Variety in practical qualities inside species is additionally key. For example, in this extraordinary component, Reich et al. feature enormous intraspecific variety in gymnosperm needle qualities with scope over the immense boreal space. They take note of how this variety fits with quality financial hypothesis and model the effects of characteristic minor departure from carbon cycling across the world's boreal timberlands. They approve the model yield against free information. Moreover, concentrate phenotypic and genotypic variety inside species crosswise over space (e.g., inside the scope of species) speaks to one essential part of practical biogeography.

Future endeavors to evaluate both intraspecific phenotypic and hereditary variety inside species' reaches ought to bring fascinating bits of knowledge into the eco-developmental drivers of species' dispersions and help connect useful environment, spatial nature, hereditary qualities, and transformative science. The subject of representing intraspecific variety in cross-species quality-based examinations has been broadly talked about throughout the previous 5 years. The significance of representing intraspecific variety at macro ecological scales, when interspecific variations are relied upon to be extensive, is still discussed.

A useful quality approach has now been quickly and broadly created in plant nature, helped by the advancement of institutionalized conventions and strategies. All the more as of late, comparative activities have been started for microorganisms and creatures. These activities will support cross-scientific categorization correlations of quality assorted variety designs from neighborhood to worldwide and clear the street of a coordinated, extensive system for the comprehension of nature everywhere spatial and fleeting scales. In this exceptional element, Whittaker et al. can give new experiences in the field of island biogeography by building useful diversity–region connections, in supplement to species richness–range connections and analyze them crosswise over scientific classifications (here amongst insects and arachnids).

The investigation of harmoniousness of quality decent varieties crosswise over gatherings is additionally opportune for better focusing on preservation territories, measuring and clarifying multitropic arranges and assessing biological community administrations. Strikingly, in this exceptional element, Kembel et al. show how phyllosphere bacterial groups affect the execution of plants and the structure of plant groups in a Neotropical wood. The creators suggest that the phyllosphere bacterial assorted variety is a key part of plant working and ought to be considered as a plant practical quality, henceforth broadening our present meaning of utilitarian attributes. However, this investigation reveals insight into the natural part of the expanded phenotype, still to a great extent overlooked in attribute-based environment; specifically, its significance for the adaption of life forms and for the associations among them.

All the more comprehensively, such spearheading examinations are a stage forward in the evaluation of species associations (counting mutualism, rivalry, help, predation), which remains a need question in group nature and biogeography. Distinguishing the characteristics engaged with species connections (association qualities from now on) is testing yet significant for a few methodologies including system examinations. Considering the associating groups (the phyllosphere bacterial groups) as attributes of different life forms (the plants) is one promising advance toward that path.

8.4.2 From Patterns to Predictions

Despite everything we need learning of the reaction of organismal characteristics to ecological changes for most kingdoms, including plants, which is, in any case, an essential to make attribute-based environment a prescient science. The main trait–condition connections were built up from nearby examinations, yet site-subordinate impacts and the limited scope of natural conditions shrouded in many investigations question the strength of the outcomes. A utilitarian biogeography way to deal with this inquiry is an applicable option. In fact, looking at how species and their functionalities shift geologically can be valuable in light of the fact that a considerable lot of similar drivers of progress that happen at each and every site on earth have effectively fluctuated crosswise over time and crosswise over space.

Furthermore, the investigation from Reich et al. is an eminent case. Utilizing a huge worldwide biomass dataset, the creators give biogeographically unequivocal connections between biomass parceling in trees and temperature, explaining a long-standing natural inquiry concerning the variety of plant biomass distribution with expanding unpleasant conditions.

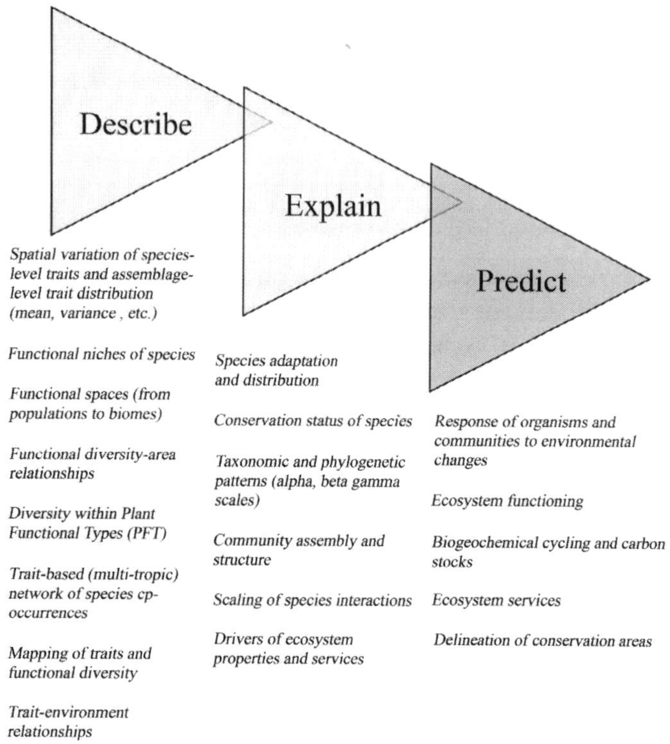

Figure 8.3: The three tenets of functional biogeography: describe, explain, predict.

Source: Proceedings of National Academy of Sciences

The foundation of robust trait–condition connections will help accomplish one center objective of functional biogeography: to put measures of the elements of biodiversity on a guide. Such practical maps are without a doubt the bases of many inquiries in utilitarian biogeography and inferred fields. In this common element, Bennie et al. richly outline center behavioral characteristic in warm blooded animals: the diel time partitioning. They demonstrate that the biogeography of this quality is affected by thermal limitations yet additionally counterfeit lighting and other anthropogenic activities. Such a finding is imperative for enhancing the demonstrating of warm-blooded animal's distribution.

On the other hand, past macro physiological implications on the biogeography of species, mapping the decent variety of organismal capacities is likewise integral to the measurement of biological system working and administrations. Here, van Bodegom et al. demonstrate that attribute information is presently accessible worldwide to delineate characteristics relating to plant working (leaf mass per region, stem-particular thickness, and seed mass in their examination) at a worldwide scale. Furthermore, they exhibit that the spatial variety of these three qualities alone clarifies a vast piece of watched vegetation sorts, in this way connecting characteristic appropriations to dynamic worldwide vegetation models (DGVMs). At last, such maps can help refine earth science and land surface models in a more persistent way, that is, by supplanting the spatial circulation of plant practical sorts by nonstop maps of utilitarian qualities as information data in those models gave that information on characteristics and species events are accessible.

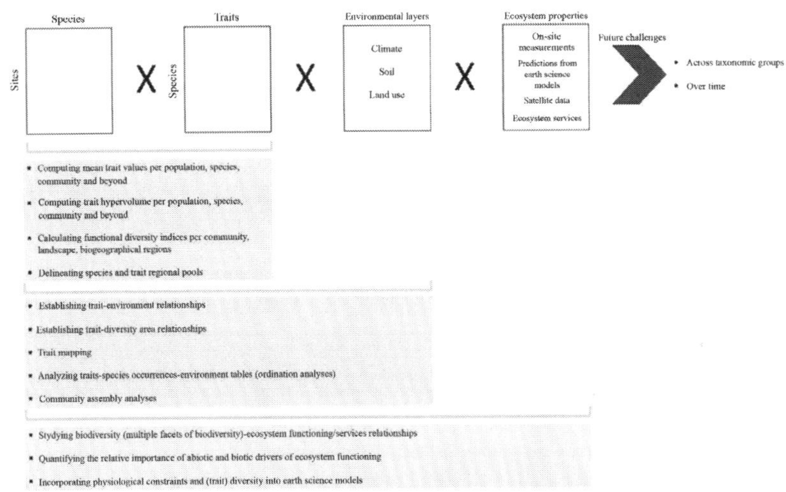

Figure 8.4: General chart for functional biogeography studies using a combination of heterogeneous data.

Sources: Proceedings *of National Academy of Sciences*

At last, the assumed connection between utilitarian characteristics and asset availabilities offers new bits of knowledge into the physiological furthest reaches of the biogeographical dissemination of species and the outline of their major and acknowledged specialties. In this issue, Stahl et al. effectively demonstrate that three key functional traits—seed mass, wood density, and plant height—explain species' range limits of a continental flora. Maps of functional traits could also be considered as new environmental layers for species distribution modeling. Such analyses are a step forward in the definition and quantification of the functional niches of species.

8.5 BIOGEOGRAPHY IN THE CHANGING WORLD

Similar scientists have either framed models that clarify a living being's history in term of phylogenetic heredities or family histories, or have thought about their perceptions as groupings. These practices of correlation and clarification are characteristic in deliberate examinations and have impacted the advancement and production of the field of concentrate known as biogeography.

There is plainly an uncommon broadness of reason in biogeography today, with the end goal that signs for the future course of concentrates in this field is befuddled, no doubt. Will tomorrow's biogeography follow the transformative histories of specific life forms case by case, or center at the level of taxa or even individual quality heredities? Should biogeography be worried about examination of biological districts or biota, or concentrate the arrangement of endemic ranges? Will biogeography be coordinated at protection or biodiversity contemplates, or will it maybe include investigation of human social orders and their communications with nature?

There are incredible numerous understandings and meanings of biogeography, each with their own particular inclination toward a specific field of study. The molecular scientists may characterize biogeography to be the investigation of quality ancestries; the biologists may demand it is about biological communities and their land extend; a systematist may support a biogeography that arranges locales and biota. Biogeography is experiencing something of a recognize emergency as the term has turned into a "catch-all" to portray the land part of a specific life-science field as opposed to existing as an autonomous field of study. It is not amazing, at that point, that the recently restored field of land phyllogenetics, called phyllogeography, which joins quality trees, has pulled in such a large number of supporters so rapidly.

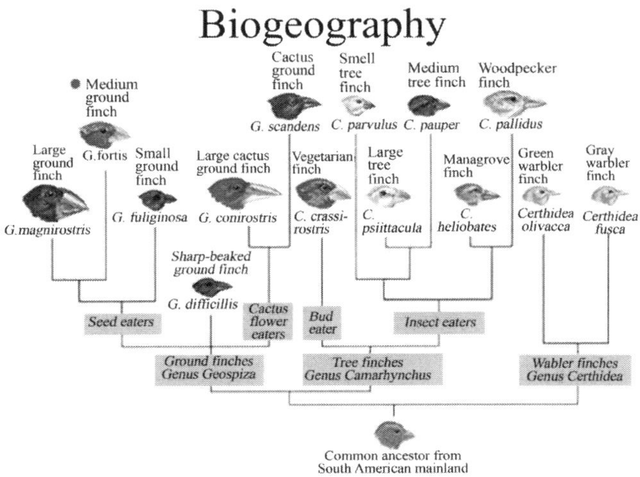

Figure 8.5: Evidence of evolution biogeography.

Biogeography, as a word, is generally new, being instituted, however not characterized by Clinton Hart Merriam. In every last case, biogeography or zoogeography or phytogeography or general biogeography is portrayed distinctively by its creators. The systematists and taxonomists rehearsing biogeography liked to portray endemic territories when all is said in done grouping. The more developmental disapproved of bio geographers, for example, Buffon and Ernst Heinrich Philipp August Haeckel (1834–1919), favored an unequivocal history of dissemination and advancement. It has been the last propensity that has the best effect on 20th-century biogeography, prove in progress in the works of scientists, palaeontologists developmental scholars and off lately, sub-atomic systematists. Biogeography has taken after a comparative pattern in its methodologies, as has natural systematics. Orderly techniques speculations and usage, for example, homology, synapomorphies, parsimony, and cladograms, have majorly affected a deliberate biogeography, which is the methodology that endeavors to arrange. A similar proof of the significance of the techniques speculations and usage applies to transformative science. The field that has supported founder dispersal and molecular time keeps has established transformative biogeography—a field that endeavors to clarify individual histories by withdrawing their land and developmental pathways.

Systematic biogeography vs. transformative biogeography is a characteristic strain that exists in view of the expectation of its experts. Truth to be told, the "two biogeographies" are similarly environmetal, verifiable, and paleontological. Contrasts between these fields mirror the splits shown in developmental science and systematics. The previous means to seek components inside phylogenies (e.g., atomic systematics and fossil science), and the last expects to give groupings of endemic zones and biota (as in sub-atomic systematics and fossil science). Moreover, a systematic and transformative approach exists in all natural, land and topographical field. In fields where progenitor relative heredities are proposed, so too are sister gatherings. Where we propose geomorphological systems, we additionally portray sedimentary structures. Where we foresee sympatric or patapatric speciation between populaces, we likewise depict populace structure. Where we decide structural elements, we additionally portray mineral morphology and the structure of rocks.

The divisions inside biogeography reveal the similar divisions that exist between all verifiable, that is non-explanatory science. What makes the investigation of biogeography one of a kind is the thought of room, to be specific in connection to topographical dissemination and assorted variety. As biogeographers we may either clarify it or group it, yet we can't do both in the meantime.

Without a doubt, the greater part of individuals who call themselves biogeographers hone developmental biogeography—as confirm by the clarifications, which seem to legitimize and transformative "reason" (dispersal components, relocation designs, recognizable proof of species-rich focuses of inception or broadening). Alternately, the individuals who seek after characterizations incorporate every single conceivable clarification, yet in addition dare to pick among them. The proceeding with struggle in biogeography isn't one of identity, philosophy, or usage it is a contradicting goal. One goal is to demonstrate that with express histories in play we needn't bother with general characterizations and on the other hand, there is the expectation to reveal widespread orders that mitigate the requirement for unequivocal case histories. In the event that we mean to confide in our models and settle on an approach to execute a technique to produce all conceivable recorded situations and pick among them, at that point we have selected a binding together strategy or approach.

However, we may, notwithstanding, choose to order and recognize all conceivable biota and frame a characterization that may find all-inclusive examples. Under an all-inclusive approach we recognize that a few histories may never be found or settled. The decision in approach is totally up the biographer—either produce models and make declarations in light of what these models foresee, or find examples and after that talk about what such pieces of our Earth's history mean.

Certain parts of biogeography, similar to those of systematics, fossil science, environment, and topography are winding up less and less near. Without a comment, similar science ends up plainly trivial, and systematics and biogeography lose their setting in an evolving world.

8.6 FUTURE PERSPECTIVES

Functional biogeography should help enhance our comprehension of the biogeographical examples of species decent variety and, at last, may enable us to foresee the outcomes of worldwide changes for biological community capacities and administrations. Nonetheless, the improvement of desires particular to biogeography is in its outset. To make functional biogeography a prescient science, it will be critical to incorporate the eco-development flow of life forms, groups, biological systems, and past. Some current discoveries propose a tight linkage between practical attributes and demography, which speaks to a promising advance in the improvement of an integrative and dynamical hypothesis to utilitarian biogeography.

Another need road is to integrate bits of knowledge into trophic or nourishment web models, first by giving maps of elements of essential makers. To do as such, it is dire to better portray qualities crosswise over scientific categorizations and their interrelationships, and to distinguish and measure collaboration characteristics. Moreover, fast advance has been made in eco-informatics; yet, the worldwide scope of practical qualities, notwithstanding for plants, is inadequate. Until the point when more noteworthy scope is accomplished, it will be difficult to accomplish a precise estimation of characteristic decent variety at a worldwide scale and to represent a few aspects of quality assorted variety, including practical irregularity.

A utilitarian point of view to biogeography can without much of a stretch incorporate intraspecific attribute variety to survey the significance of local adaptation and phenotypic plasticity in the regulation of substantial scale forms. By and by, late open deliberations in useful biology have propelled our comprehension of the significance of intraspecific variety at nearby scales, it is pressing to assess its significance at large scales. It has been however expected that species turnover and interspecific varieties prevail over intraspecific variety at large scales, except observational tests are rare and for the most part originate from the plant kingdom. Furthermore, there are some particular regions of useful biogeography where intraspecific variety ought to be represented, specifically while portraying the useful space of a given animal types and its phenotypic variety inside its appropriation range. More recently, neutral models have been used to test important mechanisms driving diversity such as dispersal patterns that helps in explaining large-scale diversity patterns.

REFERENCES

1. Cyrill, V. (2014). The emergence and promise of functional biogeography. *Proceedings of the National Academy of Sciences of the United States of America, 111*(38), 13690-13696.
2. International Biogeography Society. (2013 March). *Frontiers of Biogeography*, *5*(1). Retrieved from http://www.biogeography.org/html/fb.html
3. Luoto, M. (2006). *The potentially of remote sensing in biogeographical research.* Thule Institute, University of Oulu: Nordia Geographical Publications.
4. Marquet, P, Allen, A. P., Brown, J. H., Dunne, J. A., Enquist, B. J., Gillooly, J. F., Gowaty, P. A., Green, J. L., Harte, J., Hubbell, S. P., O'Dwyer, J., Okie, J. G., Ostling, A., Ritchie, M., Storch, D., West, G. B. (2014). On theory in ecology. *BioScience, 64*(8), 701-710. Retrieved from http://bioscience.oxfordjournals.org

INDEX

D

E

F

G

H